THE OFFICER SURVIVAL MANUAL

SECOND (REVISED) EDITION

by Devallis Rutledge

CUSTOM PUBLISHING COMPANY

Copyright 1988 by Custom Publishing Company

All rights reserved. No portion of this book may be reproduced in any manner without the prior written permission of the publisher, except for brief passages which may be quoted for book review purposes.

Library of Congress Catalog Card Number: 88-71231

ISBN 0-942728-36-X

CUSTOM PUBLISHING COMPANY
1590 Lotus Rd.
Placerville, CA 95667

Order Inquiries: 916 626-1260

THE OFFICER SURVIVAL MANUAL

SECOND (REVISED) EDITION

About the Author . . .

Devallis Rutledge is the author of:

COURTROOM SURVIVAL: The Officer's Guide to Better Testimony
CRIMINAL INTERROGATION: Law and Tactics
The *NEW* POLICE REPORT MANUAL
The OFFICER SURVIVAL MANUAL
The SEARCH AND SEIZURE HANDBOOK For Law Officers

Rutledge's books are widely used by colleges and law enforcement agencies in all fifty states and several foreign countries.

Rutledge is a former veteran police officer, holds a law degree and is currently a prosecuting attorney in Orange County, California. He has taught law enforcement subjects to hundreds of students and officers in more than fifty colleges and police, sheriff's and highway patrol departments in several states.

This book, like all of his others, is written for the student and field officer in clear straightforward language—with scores of examples taken from real-life field situations.

CUSTOM PUBLISHING COMPANY

CONTENTS

Chapter	Page
1 IT'S YOUR FUNERAL 11
Illustrating the importance of survival.
2 PERSPECTIVES ON SURVIVAL 23
What are your major controllable risks? Which risks are the biggest threats to your survival?
3 SURVIVING STRESS 29
Stressful Nature of the Work 32
Stress From Community Sources 42
Stress Within the Department 47
Stress From the Justice System 61
Stress in Family Relationships 64
Stress From Yourself 65
4 SURVIVING CARDIOVASCULAR RISKS 71
Understanding the Risk 72
High Blood Pressure 74
Diet 75
Smoking 76
Exercise and Body Composition 77
5 SURVIVING TRAFFIC RISKS 85
Routine Patrol Driving 86
Emergency and Pursuit Driving 93
Vehicle Safety 99
Safety Devices 100
Pedestrian Hazards 102

6 SURVIVING SUICIDE RISKS 107
  Who Commits Suicide? 108
  Why Do Officers Kill Themselves? 109
  The Warning Signs 111
  Suicide Prevention 115

7 SURVIVING FIELD THREATS 121
  Survival of the Fittest 122
  You Bet Your Life 123
  Motivation 127
  Protective Gear 131
  Weapons and Ammo 133
  Survival Briefing 136
  Vehicle Security 137
  Know Your Location 138
  Field Interviews 140
  Carstops (Non-Felony) 152
  Felony Carstops 171
  Robbery Calls 190
  Burglary Calls & Building Searches 196
  Disturbance Calls 205
  Deranged Persons 220
  Ambushes 221
  Plainclothes & Off-Duty Risks 230
  Arrest and Control 236
  Use of Force 259
  Firearm Retention 282
  Under The Gun 288
  Safety Fundamentals 294
  Out of Sight 299
  What If? 301
  Disguised Weapons 302
  Your Own Worst Enemy 321

8 SURVIVING WOUNDS AND INJURY 325
    Don't Panic 327
    Stop the Bleeding 328
    Protect Your Air 336
    Prevent Shock 339
    Other Self-Aid Principles 340
    Recap 346

9 IT'S YOUR LIFE 349
    Coming home alive.

Some of the photographs in this book were posed, for illustration purposes only. They are not necessarily intended to depict the tactics or procedures used by the particular officers or departments portrayed.

THANKS

I'm grateful to Chief David L. Snowden, Lieutenant Jack Calnon, Sergeant George Yezbick, Photo Lab Supervisor Philip Bettencourt, Senior Officers Robert L. Crogan and Tom Pipes and Officer Harriet J. McClain, all of the Costa Mesa, California Police Department, for their cooperation in the production of this book.

Devallis Rutledge

In Memoriam,
Officer Nelson Sasscer
Santa Ana Police Department
Santa Ana, California
1969

1

IT'S YOUR FUNERAL

In your lifetime, you've known people who died. You've heard of many others. But *you* always lived on. You survived. You're still alive.

That kind of experience tends to lead you into a false sense of security—of your own invulnerability. It makes you tend to think of death as something that continually strikes thousands and thousands of other people—both older and younger than you—and always leaves you still alive. And eventually, you begin to take your continued survival for granted.

But you *know* you're not really exempt. You could die today, tomorrow, or next Friday. What would it be like?

12/THE OFFICER SURVIVAL MANUAL

Next Friday, 4:31 p.m. You're found in the street, face down.

IT'S YOUR FUNERAL/13

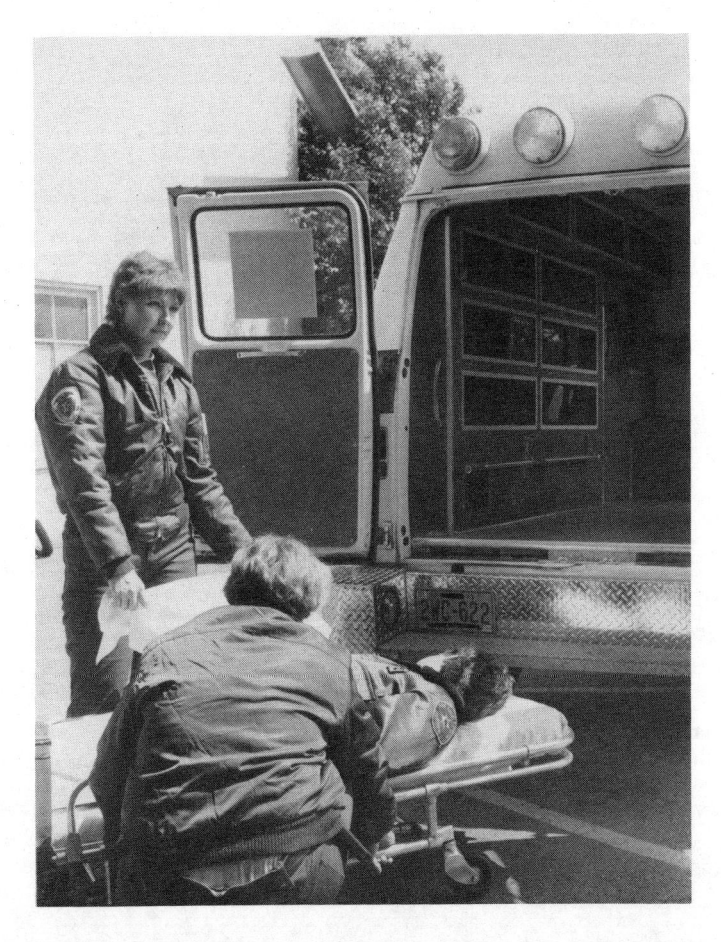

Friday, 4:44 p.m. You're loaded into an ambulance.

14/THE OFFICER SURVIVAL MANUAL

Friday, 4:53 p.m. Inside the hospital emergency room, doctors pronounce you dead on arrival.

Friday, 5:09 p.m. An officer knocks on your front door, dreading the message he has to deliver to your family.

16/THE OFFICER SURVIVAL MANUAL

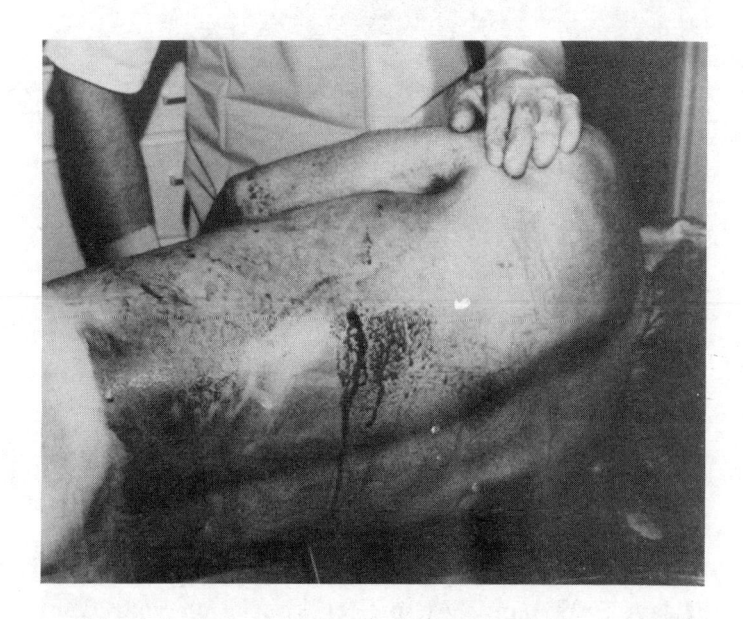

Saturday, 9:20 a.m. The medical examiner performs an autopsy on your body.

IT'S YOUR FUNERAL/17

Sunday, 2:01 p.m. The mortician prepares you for burial.

18/THE OFFICER SURVIVAL MANUAL

Monday, 10:00 a.m. Your memorial service is attended by officers from your department and the surrounding area.

IT'S YOUR FUNERAL/19

Monday, 10:35 a.m. Your casket is loaded into the hearse. Officers escort your funeral procession to the cemetery.

20/THE OFFICER SURVIVAL MANUAL

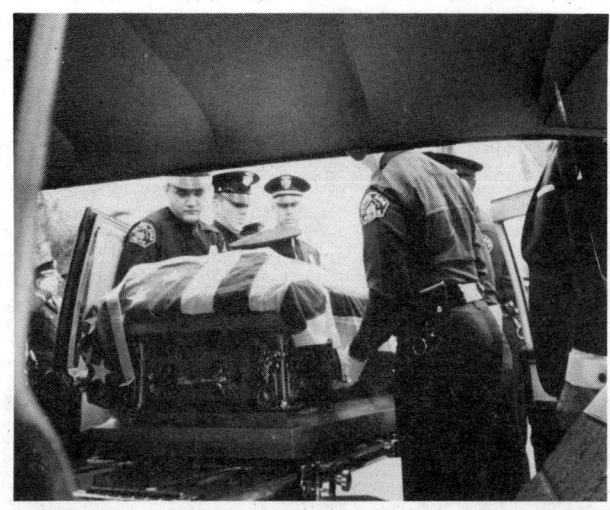

Monday, 10:54 a.m. Pall bearers move your casket from the hearse to your grave site. After a brief graveside service, officers file past your final resting place as they leave the cemetery.

IT'S YOUR FUNERAL/21

Monday, 11:23 a.m. After everyone is gone, cemetery workers lower your casket into a hole in the ground and dump 108 cubic feet of dirt in on top of you. The front loader adds a mound of earth on top.

22/THE OFFICER SURVIVAL MANUAL

Monday, 11:39 a.m. As the workers move on to their next job, your remains lie beneath a flower-covered mound of dirt. A small metal sign at the head of your grave marks the spot where next month a marble stone will carry the epitaph of an officer who died too soon. A few miles away, the personnel department will begin the process of selecting your replacement. And your family will begin the sad task of facing life without you.

2

PERSPECTIVES ON SURVIVAL

On the average, a law enforcement officer dies in this country every 3 hours and 54 minutes of causes which could probably have been prevented. By this time tomorrow, 6 more officers who are living at this moment will be dead from these same causes. The sad and terrible truth is that the average life expectancy of law enforcement officers is 14% shorter than for the general population.

What do these statistics mean for *you*, personally? They mean that *you* are in a high-risk occupation. They mean that *you* run an increased risk of dying an unnecessarily premature death. They mean that you have to become more concerned with considerations of personal survival than most other people do.

Those 6 officer funerals next Monday aren't a matter of fate—they're a matter of statistical averages. But statistics can be changed. There is no natural law that says 43 officers must die each week of controllable causes. Officers everywhere can change that statistic and drastically reduce that average.

Where do you start? Between your ears. You start by making up your mind that you're not going into the ground next Monday. You make up your mind that you're going to survive—that you're going to control your risks. You make up your mind that you're going to stop taking survival for granted—that you're not going to merely *assume* you'll always come home alive and well after the next shift, simply because you always have before. You make up your mind that you're going to set survival as a personal goal, and that you're going to work at it, just as you do with any other kind of goal.

This doesn't mean you have to become paranoid, or that you draw down on every driver you stop, or call for a back-up every time you make a move, or sleep with one eye open and your finger on the trigger. You don't. But you do have to remain consciously concerned with proper survival considerations throughout your law enforcement career, day-in and day-out. The risks never go away. Neither should your concern for survival.

IDENTIFYING THE RISKS

The first step to making yourself survival-conscious is to recognize the risks, and to put them into their proper proportions. Every officer survival publication I've ever seen has failed to do this. They zero in on the *dramatic* risks—such as shootings and stabbings—and offer you some advice on how to handle these, *after* you've gotten yourself into them. They don't spend much time suggesting ways to *avoid* these dramatic incidents, and they often neglect to tell you that shootings and stabbings are the *least* of your worries.

However, I don't recommend that you ignore such publications, because any book on survival—no matter how limited in focus—is well worth reading. You can't know too much about staying alive. If I were you, I would read every such book or article that I could get my hands on. There's no such thing as a complete book on any subject; every author will contribute something new and different, and any single tactic or tip from a different source could be just the thing to save your bacon someday.

What I *do* recommend is that you not make the same mistake that many academy instructors, criminal justice professors, and survival writers routinely make: *do not treat survival as a single-risk subject,* focusing only on hostile field encounters with an armed adversary as the only kind of threat to your survival.

The fact is that you are more likely to commit suicide than to be killed by an armed opponent. You are even *more* likely to die in a work-related traffic accident. And you are still *more* likely to die of work-related cardiovascular problems. *Survival* means surviving *all* of these risks—not just the least threatening but most dramatic of them.

If you aren't sure whether you agree with that, and if you think you're only interested in learning how to survive hostile field encounters, then check the appropriate boxes below to identify the risks you're not concerned about:

☐ *I wouldn't mind my funeral being next Monday, as long as I die of suicide.*
☐ *I wouldn't mind my funeral being next Monday, as long as I die in a work-related traffic accident.*
☐ *I wouldn't mind my funeral being next Monday, as long as I die from a work-related cardiovascular disease.*
☐ *I wouldn't mind my funeral being next Monday, as long as I die from a hostile field encounter.*

The reason you didn't check any of those boxes is because you don't want to die right away—not from *any* cause. Those 4 causes are largely preventable, and yet those 4 causes kill 43 officers a week, 187 officers a month, 2,247 officers just like you every year. *Of all officer deaths, only 2.3% result from hostile field encounters.* If you concentrate your survival instruction only on this 2.3% risk, you're ignoring most of the threat to your goal of staying alive. That's not giving yourself very good odds for survival.

If something kills you next Friday, it won't make any difference to you which way you died. It won't make any difference to your family. You'll be just as dead, no matter which way it happens. So don't take any of the threats to your survival lightly. Suicide, traffic

No matter what the cause, dead is dead.

accidents, and cardiovascular disease may not be as interesting and as dramatic to you, but the death they produce is completely indistinguishable from the death caused by a gunfight in the middle of the street. The only difference is that these 3 causes of death are easier to prevent. And they're 25 times more likely to kill you.

LESSER INCLUDED RISKS

Fatal risks aren't the only ones you face. Some very uncomfortable things short of dying can also happen to you—paralysis, loss of senses, loss of limb, non-fatal injuries and non-fatal diseases—all of these possibilities are more likely for you than death. For example, **an average of 95 officers are feloniously killed each year, at the same time, about 22,000 (1 in 16) suffer personal injuries resulting from assaults.**

For the most part, the same tactics you follow to prevent *fatal* injuries and diseases will necessarily prevent or minimize most *non-fatal* injuries and diseases. There are a few safety tips, however, scattered throughout the book, at appropriate places, which don't deal with life-threatening problems, but which may save you a finger or an eye or even a bad back. Staying healthy and intact is not as important as staying alive, but it *is* a legitimate part of your survival goal. So even if you never need to rely on any of the suggestions in this book to save your life, your precaution in following them may be responsible for preventing or lessening an injury or an illness several times during the course of your law enforcement career. □

3

SURVIVING STRESS

Surviving all 4 categories of your job-related risks (disease, accident, suicide, homicide) requires that you maintain good mental health. You can be as big and tough as they come, and you can carry a .44 Magnum, a .22 Derringer, a riot baton, a sap, a quart of mace and a loaded shotgun, and you can wear armor from head to toe, but you cannot survive for long unless you develop a realistic and healthy mental attitude that allows you to keep a clear, alert mind, free of excessive anxiety and depression, and able to cope with *all* of the various threats to your continued existence.

The connection between a poor mental outlook and *suicide* is obvious. The part poor mental fitness plays in contributing to death by *accident* or *homicide* may be less obvious; however, mental preoccupation with nagging worries or discontent can take the edge off your alertness, can prevent you from thinking clearly, can interfere with your perceptions and judgment, and can hamper your decision-making abilities. These are mental faculties which you desperately need in order to avoid or respond to accident risks and hostile field encounters. And it has been shown that an unhealthy mental disposition can cause or contribute to such *physical* problems as muscle tension, headaches, high blood pressure, peptic ulcer, low back pain, severe emotional disorders, and coronary disease, including heart attack. Poor mental adjustment can also lead to heavy smoking, overeating, alcoholism, drug dependency, sexual difficulties, and marital trouble, which can add to your suicide and coronary risks.

Since your mental outlook is directly and significantly linked to your ability to survive the job-related risks of suicide, homicide, accident and disease, *mental fitness* must be a part of your program for insuring that you meet and maintain your survival goal. If you're serious about doing everything you can to beat the odds on your premature death, then let's begin by identifying and confronting the common challenges to the law enforcement officer's peace of mind, and considering some possible approaches that may help you cope successfully.

I've grouped the stress-related factors into 6 groups for discussion; however, your mind doesn't sort these factors into groups and stick them into mental compartments. All of these factors interact with each other, so that you may take job worries home and home worries

Surviving your risks requires a clear, alert mind. If stress takes away your edge in a gunfight, you lose.

to work. Since these stresses can pop up anyplace and anytime, your methods for coping with them have to work full time, too.

STRESSFUL NATURE OF THE WORK

Every occupation has its unique stresses. When stresses are inherent in the nature of your work, there are 3 ways you can react to them: you can learn to cope with them, or you can let them take their mental and physical toll on you, or you can change occupations. The first and last of these alternatives are intelligent choices; letting job stress ruin your health is not. Therefore, if you should find yourself unable to tolerate the daily stresses that are inherent in law enforcement work, you should change assignments or occupations. It's not only your own life that you risk when you become a stress victim—it's also your fellow officers who depend on you in the field.

Fear and Danger. A part of your job is to control the behavior of others. That means conflict. It means arguments, fights, chases and shootouts. It means contact with drunks, drug-crazies, and deranged persons. It means being the potential target of snipers and ambushes. All of these risks can generate fear, and if you succumb to it, that fear can prevent you from doing your job and can become a heavy mental burden. Fear and danger are not normally a problem for most young officers, who enter the profession knowing the risks and generally considering themselves to be at least adventurous, if not fearless.

However, many people have never really been in fear for their lives, and they don't realize how awful such a fear can be, and how disconcerting its long-range effects can be, until they find themselves alone in a gunfight with a carload of bank robbers, or losing a barroom fight with a 280-lb. cop-hater sitting on their chests and choking them into unconsciousness. When the pucker-factor clamps down like an alligator and the stomach feels like there's someone in it with an industrial-strength vacuum cleaner, people who started out fearless sometimes find themselves experiencing a change of opinion (if not a change of underwear).

How do you cope? The best way is to increase your own confidence about your ability to survive. The best way to do *that* is to put into practice as many survival techniques as you can learn—from this book, from other publications, from survival instructors, and from fellow officers who have survived the various threats that you still face. Knowledge is the key to confidence, and *confidence* is the key to coping with fear and danger.

It may also help you to bear in mind that while 22,000 officers may be injured each year in assaults and another 110 killed, more than 400,000 other officers will survive without being killed or injured in a hostile encounter.

The Startle Effect. Several times during the week, most officers will suffer a startle. Usually, it's the radio call that follows the "emergency transmission" beep— a robbery in progress, officer needs assistance, man with a gun, or other such intense call. Several times a week when you suffer the startle, your body gets ready to defend itself: your autonomic nervous system is aroused, your heart rate increases, your blood pressure goes up, muscle tone increases, and adrenalin is discharged.

No matter what happens on the call itself, just sitting there in your patrol car and experiencing these effects of the startle 2 or 3 times a week can have disastrous effects on your physical and mental health, causing anything from a skin reaction to a heart attack. Undergoing startles too frequently and maintaining the "hyper" defenses over a prolonged period may result in depletion of bodily resources, a collapse of the organs, or the development of an "adaptive" disease, such as ulcers, arthritis, or cardiovascular disease.

How do you cope? The defensive reactions of your body are necessary survival mechanisms. They prepare you for the emergency. These reactions are involuntary, so you couldn't control them if you wanted to, and in your line of work, you wouldn't want to. What you *can* do is to take a long, slow, deep breath as soon as you get the startling call. This will cleanse your respiratory system of carbon dioxide and give your heart a breath of oxygen just when you need it. It will also give you a moment to "collect your thoughts," and will demonstrate that you haven't lost your head over the startle. When you see that you are cool enough to pause and take a deep cleansing breath, you'll be less susceptible to panicky reaction.

After the startle is over and the call is completed, do something to *unwind* before taking on another call.

This is a good time to take your coffee break or Code 7, or just go park in a safe place and do a couple of minutes of slow, deep breathing to regain your calm.

Good physical conditioning (discussed in the next chapter) can help you tolerate the startles. Also, confidence in your survival abilities is important here, too.

If your department has a policy, or your dispatcher has a habit, of needlessly startling the whole force before assigning the call, try to change that, if you can. When there's a silent robbery alarm at a westside bank, it's unnecessary and unwise to put a pre-announcement stress on every officer in the east, north and south by alerting everyone on the radio frequency that one or two of them are just about to get a hot call. I recommend that the dispatcher locate the units closest to the call, assign the call and back-up, and save the rest of the field force the wear and tear of an unnecessary startle.

Fellow Officer Misfortune. One of the most depressing aspects of law enforcement is finding out about, or even directly witnessing, a fellow officer's injury, fatal heart attack, suicide or murder. This kind of event is going to shock and sadden you, and may make you question your own sense of security for awhile. There's no way to avoid these natural reactions.

To cope with this stress, I suggest that you examine the situation that led to your fellow officer's death or injury to see what went wrong. Was he following the same survival practices that you would have used? Did he violate one or more survival principles? What were they? How did that contribute to his becoming a casualty?

When you've answered these questions, you're almost always going to find that the dead or injured officer was not properly survival-conscious. And while you're not going to feel like placing any blame on your fellow officer, you can at least learn from his mistakes, thereby improving your own chances of surviving a similar threat in the future. This won't alleviate your regret about your fellow officer's misfortune, but it should help you feel better about your own sense of security.

Dirty Work. Every job has its dirty work. Yours is delivering death notices to relatives, seeing little children starved and tortured, pulling mangled bodies from traffic accident wreckage, taking statements from the victims of rape and physical brutality, and coming into regular contact with dead bodies of every age, sex, and condition.

To cope, start with a realistic appraisal of man's mortality and vulnerability. None of the people whose dead bodies you handle was going to live forever. And remember that your dirty work involving death and injury is only a *fraction* of that in some other occupations. You may see a lot of burns and cuts and fractures in your job, but what about the doctors and nurses who treat these victims, and many others that you never see? You may see several dead bodies a year, but what about the coroners who perform the autopsies on these corpses and hundreds of others that you never see? Got the point? You don't need any imaginary burdens, so don't start thinking you're the only one who has to deal with pain and suffering and death. Many people contend with more of that than you do, and they survive the stress. You can, too.

It helps to distribute the load. If you tried to carry 200 lbs. of supplies on a long hike, you'd break your back. But you could easily split it up among 10 or 12 other hikers, and no one would have an intolerable burden. Learn to do the same thing with the dirty work. Field supervisors and dispatchers sometimes tend to lean too heavily on a few officers ("Give it to him . . . he's handled this kind of thing before"). That's a good way to burn out some officers while letting others rust out. Supervisors should make it a policy to distribute the dirty work as evenly as possible.

And *you* should make it a policy to redistribute your share of the dirty work, by *talking it out* of your system. Don't keep it all to yourself, and don't dump it all onto your wife or husband—split it up among several listeners. Get together with a few of the guys after work once a week and tell your war stories. Share a little bit with your spouse. Share a little bit with a close friend. Share some with your clergyman. Talk some of it over with your supervisor. Throw a little bit over onto your neighbor's lawn. Just don't try to carry everything by yourself and break your back when you can easily distribute the load simply by talking it out.

Overexposure to Evil. Your job puts you into close and regular contact with the slime of the earth (not to mention the worst behavior of the *salt* of the earth).

You earn your living by dealing with thieves, pimps, hypes, dealers, burglars, robbers, child molesters, rapists, kidnapers and killers. If you begin to feel, after awhile, that everyone in the world is rotten and dishonest, you can quickly lose interest in trying to do a good job, and you can even lose interest in living.

The key to coping with this depressing attitude is to do things to maintain your perspectives. Remind yourself that you only deal with a tiny fraction of the population. How many crooks do you see in a year? Now, what's the population in your jurisdiction? See how many law-abiding people you never have an unpleasant contact with? Those are the people you're working for. And unless you work in Washington, D.C. or San Francisco, those good people will outnumber the evil ones by at least 9 to 1.

Don't make the mistake in logic of assuming that because you only *see* evil, there must only *be* evil. Just as the mechanic only sees the broken-down cars and the vet only sees the broken-down horses, you only see the people with the broken-down morality. That doesn't mean everyone is worthless.

And if reminding yourself of these facts needs a boost, make a point of associating with someone other than criminals and cops once in awhile. Go out to dinner with a "normal" couple. Go to church. Join a civic club. Do something to put yourself in touch with the good people who comprise at least 90% of your local population.

Responsibility for Others. Some cops think that it's a terrible burden to have the responsibility of protecting other people. They begin to feel that they're carrying too much on their own shoulders, and they worry about living up to this responsibility. My advice, again, is: "Don't feel like the Lone Ranger." What about the welder who assembles tiny parts in a jumbo jet? What about the air traffic controller? What about the skyscraper architect? What about the school bus driver? And how would you like to be a neurosurgeon? Or a combat commander? Or an airline pilot? Or a paramedic?

There are few forms of human conduct which do not involve risk and responsibility—even the act of driving a 2-ton automobile through a busy intersection puts life and safety at risk and forces someone to bear the responsibility of avoiding the risked consequences. You can best meet the responsibilities of your job without excessive worry and anxiety when you have the confidence that comes from training and experience; therefore, unlike some of the other stress factors, this one should become less of a problem, the longer you remain in law enforcement.

Always a Cop. When the businessman or factory worker goes home for the day, he doesn't have to worry about doing anything for his job until the next workday begins. Most law officers are in a different situation, expected to take necessary enforcement action whenever a crime is encountered, even if off duty. When I was a cop, we were *required* to carry cuffs and weapons at all times, everywhere, even when off duty. Some officers feel that this 24-hour-a-day requirement is too demanding, and contributes to stress.

You may find this feature of your work a little less objectionable if you keep in mind that other kinds of workers also "take work home with them" (most doctors, dentists, linemen, plumbers, electricians and pipeline repairmen will handle far more off-duty calls in a year than you will). And it may help to consider whether you would want to be an unarmed witness to a supermarket robbery while off duty, or whether you would want to have the ability to do something more than just note the robber's description as he made his getaway.

After you've worked a beat or a district long enough to have made the acquaintance of the local bad buys, your peace of mind would probably suffer if you had to go out among them on your off-duty time *without* a gun and a set of cuffs. So this is another stress factor which should decrease, rather than increase, over time.

Boredom and Inactivity. A moderate level of activity is good to occupy you with something other than worry. Prolonged inactivity can be a source of stress, can lull you into dangerous complacency, and can cause you to overreact to minor violations.

It's important for supervisors to rotate officers among those assignments which involve high levels of inactivity. Whenever you are working such an assignment, do things to keep yourself occupied: watch for traffic and equipment violations, do more FI's, run out-of-state plates, stake out a possible burglary target, revisit the scene of a prior DWI arrest and make sketches for an upcoming trial, or even shake a few shop doors in the

the business district. Don't just drive around for 8 hours, listening to the radio. Keep active.

Promotional Exams. I've seen street-hardened veteran cops remain cool in the face of danger, and then start coming all unraveled when they found out the sergeant's exam had been scheduled. By the time the written and orals were over, these guys were nervous wrecks.

If the prospect of a promotional exam worries you too much, maybe you're not ready for the promotion. Think about skipping this one and waiting til the next one. Or think about taking some vacation or comp time to prepare, so that you don't create any undue risks for yourself or your partner by being preoccupied with your worries about the exam. There are people out there whose only worry is to get you before you get them. If you're more bogged down with anxiety and worry than they are, you're taking an unnecessary chance on being next Monday's funeral. That won't get you any promotion.

Shift Changes. Having to rotate shifts every few months is hard on both your physical and mental health, and it causes particular hardships on marriages and the relationship between a parent and young children. There are only 3 ways I know of to address this problem: get plenty of rest when you're off, stay in good physical shape, and try to influence your department's shift rotation policy so that it doesn't work an undue hardship on any particular officers.

If the present rotation cycle on your department is 3 months, it could easily be extended to 4 months. That way, officers would have to make the sleeping and personal adjustment only 3 times a year, instead of 4. Depending on the level of activity in your department or territory, a 6-month cycle might be feasible, reducing the adjustment frequency to twice a year.

It's not a good idea for you to stay on swing shift or graveyard permanently, just to accommodate your other job or school schedule. Medical evidence suggests that certain "biorhythms" occur in man only by day, and others only by night. Prolonged interference with these natural rhythms by living a nocturnal existence could produce health problems for you.

STRESS FROM COMMUNITY SOURCES

When law officers make so many personal sacrifices and take so many personal risks in order to provide the community with service and protection, they quite naturally feel that they deserve solid community support. And they do. But community support isn't always that solid, and this fact sometimes leads officers to feel unappreciated or even downright betrayed ("I've put my butt on the line for these people, and now they want it in a sling!"). The resulting disappointment and frustration may be sources of stress.

Public Image. There have been some notoriously corrupt law enforcement officials and departments within the past 50 years, and modern-day lawmen are still trying to rub the tarnish off the badge. Hollywood has done (and continues to do) a pretty effective hatchet job on the law enforcement profession. Television series have consistently misrepresented the lawman's role and methods. Novelists and playwrights have demonstrated a compulsive obsession with the seedy and morbid aspects of police work and the officer's personal life, scraping the bottoms of the barrels in New York, Los Angeles, and San Francisco for police stories which don't accurately reflect the realities of law enforcement work in most other parts of the country.

The result of all this over-dramatized negativity has been a widely-held, but largely-unwarranted, unfavorable public perception of the law enforcement profession. So people call you "pig." Your neighbors feel uneasy having you in the neighborhood. Kids at school give your kids a hard time because their daddy's a cop.

If you're a rural county sheriff, that means you must be a fat, ignorant, dishonest bully, because that's the Hollywood stereotype. If you're a city cop, you must be a lecherous, unstable, drunken, suicide-prone degenerate, because that's the Wambaugh stereotype. All of this negative fiction may make salable entertainment, and it may make a few people wealthy, but it comes at *your* expense. *You* have to bear the daily strain of meeting a prejudiced, misinformed public, and trying to win their understanding, cooperation, and approval for your work by dispelling their fantasies and their fears.

Officers understandably resent bearing this unfair burden. Dealing with an unreceptive public creates a stress for the honest, hard-working officer who has done nothing to deserve the public's scornful attitude.

What can be done? The public image has to be changed. Many departments have a full-time community relations officer or staff, but the problem is really too big for a single officer in each department. The chief, sheriff, commissioner, or director should charge every officer in the command as a community relations officer. You should conduct yourself in every public contact in such manner as to expose the myths about law enforcement "monsters." Your police union or protective association should regularly complain to networks and sponsors when televised programs tend to undermine good relations. You should seize every opportunity to address school classes, civic groups, church groups, work assemblies, and any other gathering where you can appear in the flesh and subdue the image of the fictional cop. The sooner these various tactics produce a more realistic public image of your profession, the sooner you and your family can stop paying the price for all the distortion. That will eliminate some stress, and might even make it easier to upgrade the pay scale.

Bad Press. This is much the same kind of problem, except that it hits closer to home—a local news story may even give *your* name. If the newspaper, radio or television reporter distorts the account of an incident, you can easily be made to look dishonest, incompetent or brutal when that's not really the case. Or an irresponsible journalist can disclose information that compromises the case or jeopardizes you or your witness.

The only solutions to this problem are to control access to information when disclosure would have damaging effects, and to improve police-press relations. If your department has a ride-along program, you may want to invite reporters to see what your work is *really* like, to help them get a more realistic and sympathetic viewpoint. If the paper or station is engaging in editorial attacks that aren't justified, it's up to the chief executive (chief, sheriff, commissioner, director, etc.) of the department to take appropriate counter-action.

Minority Attitudes. Officers who work in areas with large black or Hispanic communities sometimes get extra doses of negative public attitudes from these areas, in the form of frequent charges of brutality and racism. Even though crime statistics clearly show that members of these 2 ethnic groups commit a disproportionately high number of crimes (based on relative total populations), most people these days are afraid to acknowledge this fact, for fear of being labeled "racist." So instead, the problem gets ignored. Everyone pretends there's no particular crime problem among blacks and Hispanics, so, of course, nothing gets done to correct it.

It's only natural that this public hoax would create problems for the officer who has to "go along with the program," officially, but who has to bump heads with the undeniable realities of ghetto and barrio crime. There's little you can do about a deep-seated, national reluctance to address this problem openly and honestly. As long as you're having more frequent contact with certain groups, you should expect to receive more beefs from these groups. What you can and must do, however, is to conduct yourself properly with everyone, so that no charge of brutality or racism against *you* will ever be warranted.

Citizen Complaints. Ever since citizens found out that the department sticks letters of complaint into the officer's personnel file, they have used this device to get vengeance on the cop who wrote their citation, or busted them for DWI, or whatever. Just knowing that any citizen you come in contact with has this potential should be enough to keep you on your best behavior at all times. This makes it easy to classify the complaint letters that do come in: either they're righteous complaints (in which case you're only going to reap the stress that *you* asked for), or they're petty attempts to discredit an officer unjustifiably (in which case you should have nothing to worry about, provided you're working for a commander who stands up for his officers when they're in the right).

If your supervisor tries to bring unearned stress on you with a reference to citizen complaints, point out how many tickets you've written, and how many drunks you've arrested, *without* complaints, as evidence that you must be doing your job right, and that this one complainant is just trying to cause you some grief because he got ticketed or arrested. (If you're getting a relatively high number of complaints as compared to other officers, it's time to re-examine your own conduct.)

Lawsuits. Criminal defense lawyers have a tendency to develop a bias against the nasty old cops who are always busting their clients. They also have a tendency to look for errors on your part which might allow them to collect an extra fee and "get back at" you by filing a lawsuit. You can be sued for violating a suspect's rights, for negligence in the performance of your duties, or for intentional wrongdoing (excessive force, wrongful death, false arrest, false imprisonment, inflicting emotional distress, etc.). If you lose such a lawsuit, you can lose your money and property in settlement of the judgment, and you might be asked to resign.

The best insurance against the stress of this kind of catastrophe is to know and observe the law on your rights and obligations as an enforcement officer. Don't give the ACLU or the local attorneys anything to hang their hats on. You should also check with the department to see whether you are *personally* covered with liability insurance for such lawsuits. If not, look into the cost of buying your own policy. Knowing that an insurance policy stands between the person who's suing you and everything you own can help your peace of mind.

STRESS WITHIN THE DEPARTMENT

When Johnny Paycheck recorded "Take This Job And Shove It," he wasn't putting out a record just for cops (as a matter of fact, the song was about a factory worker). What made the record a success was the fact that from time to time, *everyone,* in *every* occupation, feels like walking up to his supervisor and making that announcement. Job dissatisfaction is almost a way of life for many workers in all professions and trades.

There are several sources of stress from within a law enforcement department which are much the same as those you might expect in almost any kind of employment. These include *poor supervision, limited opportunity for promotion and advancement, lack of voice in operational decisions, too much work for too little pay, inaccurate and disconcerting rumors,* and *the feeling that your work is unrecognized and unappreciated.*

Poor supervision is a common complaint, partly because people who really know how to supervise are *uncommon,* and partly because few organizations provide supervisors with any kind of training on how to handle authority. All you can do if you're saddled with an inept supervisor is to talk things over with him, talk things over with *his* supervisor if necessary, or just try to keep your distance until you can get a change of assignment or shift.

Limited opportunity for promotion and advancement is a fact of working life everywhere, because no business or organization needs as many supervisors as workers or troops. If everyone in the department wore sergeant's stripes or lieutenant's bars, they would have no one to supervise, and there would be no one to do the important work. You have to accept the fact that most career workers in any profession—including yours—are not going to be able to crowd into the limited number of supervisory slots that become available. (A good chief executive officer will find ways to reduce the dissatisfaction with this prospect by creating intermediate titles, positions and assignments which allow his officers some sense of progression, such as "Senior Patrol Officer," "Master Patrolman," "Lance Corporal," "Corporal," "Crime Scene Investigator," and so on.)

It is often impractical to give officers a *voice in operational decisions.* If policies were just a matter of taking a vote and letting the majority rule, there would be no need for supervisors. Organizations can't generally operate this way, and in an organization like yours, where command decisions often must be made on the spot, it is imperative to develop leadership capabilities in those designated to make decisions. A good supervisor will solicit the opinions of his subordinates on operational decisions whenever possible, but you have to realize that it isn't always possible or practicable for the supervisor to do so.

Recognition and proper *remuneration* for your work will rarely ever meet your personal expectations, no matter where you work. If you are grossly dissatisfied in this regard, discuss your dissatisfaction with your supervisor, or someone else in the chain of command. If you can't achieve satisfaction, consider applying to another agency or changing vocations, but remember that pay and recognition will probably remain below your desired levels just about anywhere.

No matter what advice I give you about departmental *rumors,* you're still probably going to listen to them and pass them on. So I won't waste a lot of space trying to convince you not to. I'll just tell you that rumors are usually distorted by the time you hear them. They usually hurt morale. If you choose to listen to rumors and help spread them around, you're asking for any stress that may bring you.

Although you share some stress factors with everyone else in the general work force, there are others arising within a law enforcement agency which are unique to your kind of work and can cause you additional stress. Most of these are matters over which you have little or no control, requiring you simply to *recognize* them, try to *understand* why they are there, and see whether you you are willing to *accept* them as conditions of your employment. (As you are promoted, you may be able to take some action to *change* some of these policies or practices.)

Inadequate Training. There's no other profession in the world that requires so many different kinds of duties as police work. All a painter has to do is paint. All a lawyer has to do is represent a client. All a taxi driver has to do is drive. But a law enforcement officer has to direct traffic, investigate accidents, investigate crimes, arrest offenders, interrogate arrestees, stop fights, break up noisy parties, quieten noisy dogs, deliver babies, give medical aid, settle disputes, guard dignitaries, write reports, testify in court, control crowds, evacuate neighborhoods, operate jails, patrol highways, find missing persons, babysit juveniles, and contend with deranged persons. A typical police academy gives instruction in *65 subjects,* crammed into a few weeks. Learning everything you need to know about science, medicine, state law, federal law, and human relations can take you several *years.*

It's not surprising, then, that officers often point to inadequate training as a source of job stress. Short of having something like a law enforcement West Point, there is no way for an academy to begin to fill in all the blanks for you. By necessity, most of your knowledge is going to come from on-the-job training, over a period of several years. Financial and manpower limitations simply don't allow for lengthy and thorough training in a basic academy.

And because most departments have to field 3 eight-hour shifts in every 24-hour day, there's no time left over for extensive in-service training. Every department, however, can make the minimal effort of providing roll-call or assembly training segments of 5 or 10 minutes, 3 times each week, to cover changes in laws, new court decisions, and enforcement techniques. Beyond that, you may be left pretty much on your own for training. There are a number of good sources for law enforcement training publications, and trade magazines such as *Police Chief, National Sheriff, Law and Order* and *Police Product News* feature regular articles on various training topics. (Your department may be hiding some of these publications in the library or chief's office. Ask to see them, or take out a personal subscription.)

Inadequate Equipment. This is another problem cited by officers as a stress-generating complaint. The most common objection is to worn-out cars. Most departments don't buy the best—they buy the cheapest. And since you normally get what you pay for, the cheapest cars don't hold up for 60,000 miles the same way better-built cars do. There's a lesson in false economy to be learned from this, but some departments have such lean budgets that they don't really have any choice. And government charters often *require* the department to accept the low bid, no matter what kind of junk this forces you to drive. The only thing I can suggest is that you invite your commander to ride in your worn-out unit sometime, so that he'll see what a piece of junk it is and will raise a louder howl the next time he's ordered to buy from the low bidder.

Other equipment complaints vary widely, but most of them tend to apply to non-critical items that are rarely or never used. Aside from personal weapons and individual protective gear, much of the exotic hardware collected by larger departments tends to collect dust in the basement. Just because you don't have some of the toys you see demonstrated on TV doesn't mean you are inadequately supplied. It's probably better to buy *quality* weapons, body armor, radios and cars than to squander part of the budget on mobile command centers, armored vehicles, and special SWAT uniforms if the department doesn't have a *regular* use for such items.

Paperwork. Cops hate paper work, and an active officer may spend 1/3 of his time doing it. The most important thing in alleviating the stress of paperwork requirements is for supervision to make certain that officers are not being asked to fill out unnecessary forms or to give unnecessary or duplicate information in any form. If you're not the supervisor, and if you don't see the need for some of the forms or some of the questions you're required to answer, ask your supervisor about it. Make him justify the paperwork, and ask him to explain why you have to duplicate the same information in different places. This will either reveal the explanations to *you,* or it will reveal the lack of them to your *supervisor,* which hopefully will prompt a streamlining of the paperwork. But since you work for a governmental agency, you're never going to eliminate paperwork altogether. Try to accept it as a part of your job.

No Follow-up. The beauty of working in a small department is that you can usually handle your cases from start to finish. You get to see how they come out. In mid-sized and larger departments, cases may originate with a patrol officer, then be turned over to the detectives for further work, and then be turned over to the criminalist for completion and then go to a large prosecutor's office for trial. The patrol officer who started the case often never hears anything more about it after he turns it over to the detectives. This gives his work a dangling, incomplete aspect and deprives him of the reinforcement of seeing a successful conclusion to the case. This lack of reinforcement can translate into feelings of futility and purposelessness, contributing to depression.

The solutions are obvious. If you are interested in the results of your cases, follow them up with the detectives and the prosecutors to find out what happened. Put a little note on the paperwork you send up, giving your name and extension, and asking the dicks to let you know the outcome. The other solution is for detectives to take it upon themselves to give patrol officers a little feedback. A word of recognition and a report on the conclusion of a case where a patrol officer has made a good collar and done some good police work is a positive, reinforcing practice which detective bureau commanders should require, or at least encourage, in their commands.

Push For a Degree. Some departments now require a college degree of their applicants. Others require 2 years of college and expect the officer to finish degree requirements within 5 years. Others give points for degrees in promotional tallies. There is a national push for higher educational levels for officers as one means of upgrading professionalism. The pressure on officers to earn degrees prompts many of them to use half of their off-duty time studying at the nearest college. This cuts into rest time, recreation time, and family association time, contributing to physical problems, as well as heightened levels of stress.

If your department puts a premium on a degree and you can see that it's necessary for your career advancement, give yourself realistic goals and timetables for getting there. If you're inclined to attempt more than 6 credit hours at a time, talk it over with someone who's been that route, and check with the department to see how your shift and assignment schedule can be arranged to accommodate school.

Shooting Policy. Officers often disagree with the department's shooting policy, feeling that unwise restrictions have been placed on field personnel by administrators who don't have to face the dangers and who are unjustifiably limiting the officers who do, just to appease law enforcement antagonists. State statutes spell out the authority of officers to use deadly force, but some departments have imposed even greater restrictions (in one California case, it was held that an officer could be sued by a criminal for violating departmental shooting policy, even though the officer's action was *completely lawful* under the statute on use of deadly force).

If you have trouble accepting your department's policy and there isn't any prospect of getting the policy changed, you may simply have to move to a department where policies are more in line with your personal view. Restricted shooting policies and adverse publicity on police shootings have led an increasing number of officers to report that they hesitated to shoot, even when confronted with a legitimate shooting situation. Some of these officers have suffered wounds because of their hesitation, and at least one officer in a recent statement blamed such hesitation on his part for the death of a fellow officer.

My personal view is that this is an unacceptably dangerous state of affairs, creating unwarranted survival risks for the officer who does not have a clear understanding of when to shoot. From an officer survival standpoint, I cannot recommend hesitation and restraint in the face of an obvious threat to your life. Everyone is aware that newspapers merely file a brief story when a *criminal* kills an *officer,* but editorialize endlessly and engage in lengthy "investigative reporting" whenever an *officer* kills a *criminal.* Anytime the fear of being crucified by the press and the ACLU overcomes your fear of being shot and killed by an armed adversary, I think you've lost your survival perspective, and a change of vocation is probably in order. You can't take any consolation in the fact that you saved the department a lot of heat from the press if they cover you up in a hole next Monday morning.

Policy on Use of Force. This is the same problem, except that the consequences are not as severe, and so the stress from troublesome policies here is not as great. The department is financially liable if one of its officers uses excessive force. This gives the department a legitimate interest in setting guidelines for use of force. However, if the department is hiring officers whose judgment and discretion cannot be trusted without imposing detailed and overly-restrictive guidelines on force, the problem is in hiring and training, and changes should be made *there,* rather than piling on more and more layers of rules for the field officer.

Reasonable policies on use of force are for *your* benefit, too. As long as they create no inordinate risks, you should be willing to abide by them. Discuss any dissatisfaction with your commanders.

Pursuit Policy. In the old days, if someone split on you, you chased him until one of you quit, no matter how fast and how far this took you. A rising number of fatal and other serious traffic collisions, together with accompanying lawsuits, has brought some changes in pursuit policies. Some departments now specify that any pursuit above a certain speed (e.g., 50 MPH) can only be maintained if a certain kind of suspect is being pursued (dangerous felon, for example).

Cops don't like to lose. Law agencies purposely look for and hire competitive individuals who like to win. When these agencies turn around and tell their officers to let the fleeing suspect win if the violation is a traffic offense or a misdemeanor, a policy is created which conflicts head-on with the officer's personal makeup: more stress.

There's no fingertip solution for this problem, either. You know and I know that someone who splits from a $10 traffic ticket and risks his life at 120 MPH is more than likely driving a stolen car or running from a fugitive warrant. But the judges who decide lawsuits don't know this. Neither do the lawyers. And the police administrators who have to respond to these lawsuits may have trouble remembering, or selling the point. The only help I can offer is some advice on how to corner the suspect without giving him the opportunity to split. This advice is in Chapter 5, under the topic of pursuit driving.

Recruiting Policy. In order to accommodate female applicants, some departments have lowered height, weight, strength and agility requirements for employment. This allows smaller and weaker women *and men* to become members of the field force, and this antagonizes some of the bigger and stronger officers who feel that inferior specimens are being assigned to jobs they're not competent to handle.

(The rush to relax standards probably assumed its most extreme proportions when a major city police department, finding that hardly any of its female applicants could scale a six-foot fence, dropped that requirement entirely. I couldn't count the times I've chased runaway juvies and residential burglars over backyard fences. What that one city plans to do about cops who can't keep up with the opposition, I have no idea.)

I've talked with male officers from various departments who are satisfied with, and even proud of, their female field officers. Other male officers express resentment that female officers are never assigned to hot calls, are never first to arrive at potential trouble spots, are not able to keep up in foot chases, and never attempt an arrest without calling for a male backup. A department which permits any such special treatment of female officers is inviting dissension among its male officers, and is casting an unjustified veil of suspicion over *all* its female officers, including those who may be performing their duties every bit as capably and as fearlessly as the men.

Law enforcement is a dangerous profession. It's especially dangerous for someone who lacks the physical skills or the courage to handle the challenges. No department is doing any favor for the man or woman who doesn't measure up by allowing him or her to pin on a badge and place himself, herself, or fellow officers in needless jeopardy. If there is a genuine, risk-creating breakdown of your department's standards, you should, in the interest of your own survival, make your objections known at appropriate levels.

On the other hand, if you are a male officer who resents the entry of *qualified* female officers into the law enforcement profession, you will save yourself some stress by changing your attitudes as quickly as possible. Under the compulsion of equal employment opportunity laws and changing social attitudes, there is no question that qualified female applicants are entitled to work as law enforcement officers. Despite the fairly high attrition rate females suffer in academies and initial assignments, you can fully expect to continue seeing and working with female officers in increasing numbers. You can also expect to see smaller, weaker men coming in under relaxed employment standards.

Corruption. With less and less frequency, fortunately, officers are placed into stressful conflicts by the dilemma of seeing a fellow officer or a supervisor engaging in misconduct, but not wanting to play the role of "informer" on a brother officer. The best way to handle this conflict, in my opinion (and I realize it's easier said than done), is to take it up directly and confidentially with the officer or supervisor involved. Making your disapproval known in this manner may not put a stop to the officer's misconduct, but it probably will put a stop to his doing it in your presence (he probably will not risk offending you to the point of provoking you into taking your complaints over his head). This should diminish your stressful dilemma, and if sufficient numbers of other officers take a similar approach, the errant officer may be forced to abandon his bad habit altogether.

If the misconduct is something that puts lives in danger (partner has a drinking or drug problem), don't fool around with it. Likewise, if it's a matter of serious violation of ethics which could bring you and everyone else in the department into discredit, don't lose sleep worrying what to do: in *either* of these events, make a prompt report to the appropriate command level. You can't afford to be a "good ole buddy" to someone who's putting your life or reputation in jeopardy.

Internal Suspicion. Formation of "internal affairs" units in some departments is seen by line officers as a "sell-out" by a compromising chief executive who may be under pressure by advocates of civilian review boards. Supervisory personnel, on the other hand, may view such units as necessary tools to permit the department to clean up its own backyard. Whatever their actual validity may be, internal affairs units can easily lead to divisions within a department, creating a source of suspicion, resentment, and even paranoia among officers who wonder whether every fellow officer they come in touch with is a "stoolie" for the IAD. In a profession where survival may depend on close, confident cooperation among officers, this is not a desirable situation.

If you have accepted employment with an agency which uses an internal affairs unit to handle the disciplinary responsibilities which were traditionally discharged by commanders, you have agreed to that condition of employment. The best route now for minimizing stress from this source is to simply be sure you are following departmental policies and using good judgment in your individual conduct. As long as you do nothing which would interest an IAD unit, you have no need to worry.

STRESS FROM THE JUSTICE SYSTEM

Law enforcement officers are put to considerable personal risk in order to bring offenders to justice. When an officer sees the benevolent treatment accorded to criminals by the justice system, and compares that to his *own* treatment by judges and lawyers, he often finds a legitimate basis for resentment.

Court Decisions. You live in the criminal's world, witnessing firsthand the dishonesty, depravity and brutality of the street crook, as well as the loss, pain and suffering of the victim. Crime is real to you. It has a face-to-face impact on you. You have no trouble seeing the need for swift and certain punishment.

Judges and lawyers, on the other hand, live in a lawbook world, insulated from the sights, sounds and smells of crime, only reading about it in an antiseptic courtroom. They don't get the same impact. They don't have the same point of view. They tend to take a more *theoretical* (translate *unrealistic*) approach to crime and punishment. Result: court decisions and practices which may make you wonder whose side the courts are on, and may make you question whether the results you get in court are worth the risks you take in the field.

The best defense against adverse court decisions in your cases is for you to know and follow the current law on arrests, search and seizure, interrogation, and evidence collection, and to know how to write a good report and to become a good witness. The less reason you give the courts to suppress evidence or dismiss charges in your arrests, the less you will feel personally affected by restrictive court decisions.

Defense Attacks. It's the defense attorney's job to try to discredit your work on a case, your report, and your testimony, if he can. Having to contend with these attacks on you can create more stress than it should if you've given the attorney some unwitting assistance. The remedy is the same as for avoiding unfavorable court decisions—know what you're doing, and do it right. Shut the defense lawyer out with your competence.

The Prosecutor. Cops and prosecutors don't always see eye-to-eye. The prosecutor may refuse to file your case, or may plea bargain it away, or may do a shoddy job of presenting the case at trial, resulting in an acquittal. To reduce these possibilities, be sure you give the prosecutor the best case you can. Anticipate the defenses and look for the holes in your case; then do everything you can to cover the holes before presenting the case to the prosecutor. If you think he's doing an inadequate job of evaluating or handling your cases, report this to your supervisor and ask your chief to discuss the matter with the prosecutor's chief.

Court Scheduling. Court isn't usually a problem when you're working a day shift. The other 2/3 of the time, however, it cuts into your sleep time, your recreation time, your personal business, and your off-duty family life. Many times, after you spend a couple of days waiting around the hall, the prosecutor comes out and tells you he's accepted a plea bargain and won't need you after all.

The best means for reducing the waste and inconvenience of court conflicts is an on-call system which allows you to continue with your off-duty activities, subject to being called in on appropriate notice (depending on distances), once the prosecutor is *certain* your testimony will be needed that day. If you anticipate spending an extended amount of time on the witness stand (e.g., 8 days on a homicide case), talk to your supervisor about a temporary transfer to the day shift during the trial. Don't try to spend 8 hours in court and 8 hours on the street day after day—you won't be effective in either place.

Sentencing, Probation and Parole. There's little disagreement among law enforcement officers that the punishment phase of the justice process is a general failure—sentences are too light to fit the crimes, probation and parole are too quickly granted and too loosely supervised, and criminals take advantage of lenient, corrupt or naive judges and probation and parole authorities.

The best reaction to this problem is your steady, vocal protest against it. If you just quietly steam about punishment failure, the responsible officials may take your silence as a sign that you agree with, or at least are willing to acquiesce in, their policies. Instead of silently reinforcing poor policies, you can at least vent your disapproval (and some stressful pressure) by expressing your point of view in appropriate ways.

STRESS IN FAMILY RELATIONSHIPS

A number of disadvantages accrue to your family life because your work schedule often keeps you away from your spouse and children at night, and brings you in touch with them only as you pass at the front door. Your wife may be afraid to stay alone at night; she may resent the sacrifices to her social life; she may resent having to raise the kids with an absentee father; she may be suspicious about your going on late-night patrol with a female partner (most of these same problems apply if you are a married female officer).

Your children may have trouble with their playmates because their parent is a cop; your children may resent not having a "normal" father (or mother) around in the evenings and on weekends, like all the other kids do; and your family may think you treat them like "suspects" in their own home.

The result of these family strains is evident in the notoriously high divorce rates among law enforcement families. The only aid I can offer is a suggestion that you lay all the cards on the table for your family and ask for their understanding and support. It may help your spouse to talk to other law enforcement wives (husbands) about common problems and coping techniques. And it will help if you share as much of your off-duty time as possible with your family. Don't let them get the feeling that your job comes first.

STRESS FROM YOURSELF

Some of the pressures you feel are self-created and amplified. Since the stresses in this category are under your personal control, they are the easiest ones to reduce or eliminate, usually with just the adoption of a particular mental attitude.

Proving Yourself. Most people are highly influenced by what their co-workers think of them, and they try to conduct themselves so as to gain approval in their co-workers' eyes. To a certain extent, this kind of motivation is okay. But when it degenerates into a form of macho competition, requiring you to show your "guts" in foolhardy or reckless ways, it becomes a survival risk in more ways than one: the officer who lacks self-confidence and feeds excessively on peer acceptance suffers unnecessarily high levels of stress, thereby making himself more prone to physical and mental suffering; and he is more likely to take unnecessary risks in such activities as traffic pursuits, handling hot calls, and hostile field encounters; when he doesn't feel he has measured up and met with peer approval, he is more likely to become angry with himself and embark on the kind of self-destructive course that can eventually lead to suicide.

Female and minority officers often feel under an extraordinary pressure to prove themselves to other officers and to supervisors, if not to the community at large. This pressure may interfere with their ability to develop natural, relaxed relationships with fellow officers, thereby compounding their stresses, and it may tempt them into taking careless risks just for the sake of showing that they aren't afraid of danger.

You don't prove yourself by being the guest of honor at next Monday's funeral. You don't do it by coming on like gangbusters when some other kind of response is what's really needed. You prove yourself by exercising good judgment. If good judgment dictates calmness, restraint, backing off, or even retreating, you prove how good an officer you are by taking your own best advice, rather than by taking unnecessary risks. When you try to play Superman or Wonder Woman, all you prove is that you lack the maturity and the sound judgment to wear a badge and carry a gun.

If you're new to the department, and especially if you're breaking ground as a female or minority officer, don't even try to win immediate peer recognition from veteran officers. Give yourself plenty of time for that. Acceptance will come when you show them that you realize you've got a lot to learn, and that you aren't so insecure that you try to become an overnight sensation as the John Wayne of the force.

Personal Code vs. The Law. Some officers experience stress as a result of conflicts between their own personal codes of right and wrong and what the law says. You can't smoke marijuana when you're off duty and bust citizens for the same thing when you go back on without feelings of hypocrisy and guilt. The same goes for drunk driving, wife beating, or child abuse. If you're not prepared to hold yourself to the same standards set by the law, you don't have any business being the one to enforce it.

And if you have personal convictions that are contrary to the law (maybe you think marijuana and prostitution and gambling should be legal), it's important for you to realize that your role is law *enforcer,* not law *maker.* You're like the builder who is expected to follow the architect's blueprints faithfully, even if he might have designed things differently himself. Your personal code should be such as to permit you to enforce *all* laws objectively and conscientiously. If it will not, the stress this conflict will generate for you will only be relieved when you change occupations.

Challenges to Integrity. Because you have a considerable degree of personal discretion as to whether or not to issue a citation or make an arrest, you're subject to a variety of attempts to persuade you to exercise your discretion against citing or arresting. Male officers are sometimes confronted with alluring female drivers or suspects who suggest romantic or sexual favors in exchange for a break. All officers are subject to bribery attempts and citizens' threats of political influence with a judge or the chief.

All of these challenges to your integrity *can* constitute stressful conflicts; however, they shouldn't. As long as you approach your duties with the conviction that your integrity is not for sale, you should have little difficulty in handling these challenges without undergoing any stress. In fact, when someone makes an improper offer or threat to influence your official performance, you should be able to react in such a way as to bring some extra stress onto your challenger.

False Self-Blame. You may be tempted to shoulder undeserved blame for some of the things that happen to people just because you were there to do your job. For example, you see a righteous moving violation and issue a citation to the driver, who tells you this ticket is going to cost him his insurance and his license, leave him unable to get to work and back, and hurt his wife and little children if he loses his job. Or you arrest a widowed mother of 3 small children on bad check charges, and she tells a similar hard-luck story. Or you bust a man for a homosexual act in public, and he tells you his marriage, his children's happiness, his security clearance and his job will all go down the tubes because of your arresting him.

In cases like these, you may be inclined to say to yourself: "If I hadn't done what I did, all these terrible consequences wouldn't have come about; therefore, it's my fault." This kind of mistaken conclusion can load you down with unwarranted feelings of guilt and stress.

The foreseeable consequences of a violator's conduct are *his* fault—not yours. He was willing to risk all of the adverse consequences of his illegal act, and knowing the risk, he willingly jeopardized his personal, familial, and occupational situation. If he had not committed the offense, he would not have made your action necessary. You can't start taking the blame for everyone else's conscious, illegal behavior. For one thing, you can't tolerate the burden; for another, you would simply be *wrong* to conclude that you have any blame to bear.

Difficulty Unwinding. Because your job demands intense levels of alertness, you may end the shift and the work week at a "hyper" stage and have trouble relaxing. Relaxation and diversion (getting your mind off police work) are essential to maintaining a balanced mental outlook. Simple devices can help you here: a few slow, deep breaths when you feel tense will help you relax, as will physical exercise, reading (not police books), music, meditation, handcrafts, hobbies, sports, household repairs, auto repairs—even shopping. Anything which forces you to get involved in non-police activity will lower your hyper state and make it possible to relax.

* * *

I've described and categorized 43 different stress-producing factors for the law officer. I didn't lay out any magical cures for these problems, because there are none. If it were easy to eliminate these stress sources from the officer's job and personal situation, it would already have been done. These long-standing stress factors continue to confront law officers of each generation because the basic human conflicts and inherent conditions which cause them to exist are not easily removed.

I have tried to suggest helpful ways to cope with some of your job stresses. But the greatest value of this chapter should be in getting you to *recognize* and *face up* to the various stresses which may be bombarding you, perhaps without your having previously and consciously identified them. Those are the important first steps to surviving your job stress.

And if some of my suggestions for coping with particular stress factors are impractical or inapplicable in your situation, remember to make use of these general guidelines: get regular rest and relaxation; let off steam; talk things over with co-workers, supervisors, friends and family; before things get too much to handle, consider changing jobs, or at least seek professional counselling.

Don't get too depressed at this point in the book just because we've spent 48 pages listing all the things that can cause you stress. Good physical health and conditioning, together with the confidence of knowing how to avoid and deal with the dangers of your job, will go a long way toward enabling you to cope with the stress of law enforcement work. By the time you finish the remainder of the book, you're going to be well on your way to achieving this healthy, confident status. *You can survive* stress! □

4

SURVIVING CARDIOVASCULAR RISKS

The biggest single reason why law enforcement officers have a substantially shorter life span than most other people is their high incidence of cardiovascular diseases. Such problems as heart attack, high blood pressure, stroke and rheumatic heart disease make cardiovascular disease the leading cause of death in America, claiming more lives each year *than all other causes of death combined.* Studies have shown policemen—especially those in middle age—to be at greater risk for coronary heart disease than the general population. An estimated 2035 officers lose their lives each year to cardiovascular diseases (that's more than 21 times the number who are feloniously slain). Many of these deaths could be prevented.

Insuring your survival goal against this biggest of all risks may require a few changes in your lifestyle. But *ignoring* the risk could mean one *final* change in your lifestyle that would be the subject of some slow walking and sad talking next Monday. Given that choice, a few beneficial changes for your living patterns shouldn't be too hard to take.

UNDERSTANDING THE RISK

Many people take a fatalistic view of death by disease, assuming that it can't be prevented. "When your number's up, you're going, and there's nothing you can do about it," they say. And while it's true that inherited tendencies to develop a disease may sometimes be a factor, you can still improve your prospects of survival by taking prudent precautions to reduce the risk of developing or succumbing to the disease.

The first step to take is to define the problem and make sure you understand it. Once you do that, you can see that survival isn't just a matter of luck and heredity, but that it's largely a product of your personal choices and habits.

What causes most of the cardiovascular problems? A condition that doctors call "atherosclerosis." When you have an excess of cholesterol and other fatty materials in your bloodstream, they become embedded in the inner walls of the arteries. These fatty deposits build up over many years, gradually narrowing the opening in the artery and creating a rough surface on the inside walls of the artery. You might compare this to the build-up of lime and calcium deposits you find inside a household water pipe after a few years of use.

Lengthwise view and cross-section of a healthy, unobstructed artery.

Lengthwise view and cross-section of an artery showing atherosclerosis. When fatty deposits obstruct your arteries like this, you're a prime candidate for deadly strokes and heart attacks.

The roughened surface of the artery walls can cause a blood clot to form. As the clot gets larger, it blocks the narrowed artery and halts the flow of blood. If this clotting occurs in a vessel serving the brain, you have what's called a "cerebral thrombosis." The blood supply to a part of your brain is cut off, resulting in a *stroke* that can leave you paralyzed, cause loss of memory, impair your vision and speech, or kill you. Strokes kill nearly 200,000 Americans every year.

If the clotting occurs in one of the arteries supplying blood to the heart, the blockage is called "coronary thrombosis." If the narrowing and clotting are severe, you can have a heart attack. Heart attacks kill more than 630,000 Americans per year.

Obviously, the longer you've been collecting excess cholesterol and fat in your circulatory system, the more obstructed your arteries will become, and the greater your risk of stroke or heart attack. That explains why the risk of cardiovascular death increases with age. This does *not* mean, however, that officers in their 20's and 30's are safe from the risks—just that their odds are a little better for a few more years.

Fatty deposits causing some degree of atherosclerosis have been found during autopsies of children and teenagers. And when examinations were performed on the bodies of soldiers killed in Korea, the majority of them were found to already have some degree of coronary atherosclerosis. Their average age: 22.

What can you do to reduce your own risk of cardiovascular death? While there's no guaranteed course of prevention, the American Heart Association has identified several high-risk factors which you can do something to control.

HIGH BLOOD PRESSURE

Untreated high blood pressure, or hypertension, can cause damage to your heart, kidneys and nervous system. Having high blood pressure increases your risk of heart attack, stroke or kidney failure.

Although the causes of high blood pressure are not all known, a heightened level of emotional *stress* will elevate the blood pressure level. It is important, therefore, to do what you can to control the stress in your daily life, as we discussed in the last chapter.

You may have high blood pressure without knowing it (an estimated 35,000,000 Americans have it), because there are no visible symptoms in many cases. The only way to find out is to have your blood pressure checked regularly by your doctor. If yours is abnormally high, the doctor will prescribe a diet and medication to control it.

DIET

One way to reduce the excess cholesterol and fats in your arteries is to reduce the amounts of these substances which you eat. The foods that are high in cholesterol are egg yolks, liver, kidney, sweetbread, and shellfish, such as shrimp. The AHA recommends that you eat no more than 3 egg yolks a week, including those that are mixed with other foods, such as pancakes, french toast, cake, brownies, noodles and mayonnaise. It is also recommended that you limit your use of shrimp and organ meats.

Saturated fats—the kind that can ruin your arteries—are found in red meat, butter, cheese, cream, whole milk, lard, and chocolate. To replace these with *polyunsaturated* fats—which tend to lower blood cholesterol levels—eat more poultry and fish in place of meat, use low-fat milk and cheese, and use vegetable margarine and cooking oils instead of butter and lard.

The AHA also suggests that you avoid fried foods and restrict your intake of luncheon meats, sausages and salami, and cut down on sugar and caffeine.

If you brown-bag your lunch, prefer a peanut butter sandwich, a chicken salad sandwich, or a tuna salad sandwich to one made with bologna or cheese. Instead of a Twinkie, take a piece of fruit. Instead of a sugary soft drink or high-caffeine coffee, drink vegetable or fruit juices (without added sugar).

When you eat out, order your pizza with mushrooms and olives, rather than pepperoni and sausage; eat a bean burrito instead of a beef burrito; order a fish sandwich instead of a cheeseburger. To improve your digestion and elimination, most nutritionists recommend high-fiber foods, instead of refined products: use whole grain bread instead of white bread; eat whole grain cereals without refined sugar; and include fruits and fresh green and yellow vegetables in your daily diet.

You can harm yourself if you make drastic changes in your diet or eliminate essential foods. Your body needs certain kinds and amounts of proteins, fats and carbohydrates, and you need to make sure that you consume calories that provide *nutrition,* instead of excess fats and sugars. Have your doctor advise you as to the kinds and quantities of foods that are best for *your* body, *your* age, *your* activity level, and *your* physical condition.

SMOKING

Cigarette smokers have more frequent and severe cases of coronary atherosclerosis than nonsmokers. Smokers have 70% higher death rates, and heart attacks are the biggest contributor to the excess death rates among smokers.

Studies have shown that people who give up smoking have lower death rates than those who continue. And after a few years, the death rates of those who quit are nearly as low as for those who have never smoked.

If you thought the risk of lung cancer was the only chance you were taking by smoking, now you know differently. You can also improve your odds of surviving cardiovascular disease (as well as many other diseases) if you kick the habit.

EXERCISE AND BODY COMPOSITION

No survival book is complete without a discussion of the importance of physical fitness. That's because *being in good condition is the single best survival insurance you can have.* It's vitally important to surviving all 4 of your controllable risks. I've chosen to discuss it in this particular chapter, however, because the risk against which physical fitness is most important and most effective is the risk of cardiovascular disease.

You may have heard of several different studies which compared prison inmates with law enforcement officers, and found the inmates to be in superior physical condition. You may also have heard that officers over age 30 are in worse physical condition than the general population in the same age group. So when someone describes a group of officers as being "New York's finest," or "LA's finest," or "Bingham County's finest," for example, that's probably an inappropriate description.

The truth is that while applicants for employment with a law enforcement agency may be put through strenuous physical exams and agility tests, once the applicant is hired and completes academy training, both he and the department are likely to neglect to maintain a standard of fitness and a program to meet that standard. Recently, the International Association of Chiefs of Police made a random survey of 302 agencies, to determine whether improving the physical fitness of officers was a high-priority training objective of departments. Only 43 of those agencies—less than 15%—reported having any kind of physical fitness programs for their officers. Apparently, fitness is a low priority.

This book is about *your* survival—saving *your* life. It's obvious from the lack of enforced commitment to physical fitness that *you* are going to have to be responsible for maintaining your own physical conditioning. To change the odds of your dying 14% sooner than other people, you should make physical fitness the cornerstone of your personal survival program. It's up to *you* to make that commitment and to follow through.

Your Body's Exercise Needs. As you may know, your muscles, tendons, ligaments and joints tend to atrophy, or waste away, if you don't use them regularly. The ability of your respiratory and circulatory systems to transport and use oxygen diminishes if you don't regularly challenge them with vigorous exercise. So you have 2 kinds of exercise needs: improving *motor ability* with a variety of exercises to increase strength, agility, flexibility and endurance, and developing greater *cardiorespiratory fitness*.

There is no single exercise that will satisfy all your exercise needs. Don't make the mistake of assuming that all you need to do is work out with weights, or jog, or do isometrics in your patrol car. Jogging will improve your cardiorespiratory fitness, but it won't improve your strength. Weightlifting will improve your strength, but it won't do much for cardiorespiratory fitness. Isometrics develop strength, but not endurance. Calisthenics promote flexibility and agility, but not strength, endurance or cardiorespiratory fitness.

So what's the answer? Obviously, you need an exercise program that alternates stretching exercises for flexibility, games for agility, isotonic exercises for strength and endurance, and aerobic activity for cardiovascular fitness. You shouldn't try to combine all of these activities into a daily regimen, because you need a "day off" between isotonic workouts to allow muscle proteins to be built up and to remove waste products. Considering the fact that you probably don't have a lot of spare time for exercising and that motivation and self-discipline may be a problem, choose a plan that you can stick to —not one that becomes a nuisance after the first week.

I'm purposely not listing or recommending any particular types of exercise, frequency or duration for you. That's because an exercise program, like a diet, should be personalized for you by your physician, on the basis of *your* age, sex, weight, body build, medical condition, and medical history. It's possible for an apparently healthy officer to have atherosclerosis of the coronary arteries, for instance, without knowing it, in which case a spurt of sudden and strenuous activity (for example from self-planned exercise or a foot chase of a suspect) can bring on a heart attack.

I recently read of a 28-year-old policeman who signed up for intramural volleyball; to get in shape, he decided to go run around the park. He finished his run and was walking back toward his car when he collapsed on the ground. He was DOA at a nearby hospital from a heart attack.

If you're not accustomed to regular, vigorous physical activity, don't prescribe your own exercise program. Let the doctor do it. If you can't get an appointment for a couple of weeks and you're anxious to start putting your body back into shape, take daily *walks* until you can get in to see the doctor. Start moderately and build up to a brisk pace over a week's time. Beyond that, treat a prescription for exercise the same way you'd treat any other prescription: use only as directed by the doctor.

Body Composition. Middle-aged men who are significantly overweight run a 200% higher risk of suffering a fatal heart attack than other men of the same age who are of normal weight. There's only one way to reduce body weight, and that's by burning off more calories than you take in through your mouth. When you're using more calories than you're eating or drinking, your body has to obtain the additional energy by burning off some of your mass.

Many people mistakenly think that the simple answer to weight maintenance is a diet, which will force the body to convert fat for energy. Although it's true that dieting alone will reduce body weight, 65% of the loss will be from catabolized *muscle,* and only 35% from body *fat.* The net result of a diet *which is not accompanied by a proper exercise routine* will be a much greater percent loss of lean body tissue than of fat tissue. That's not a desirable result.

What you want to achieve in your weight maintenance program is an alteration of body <u>composition</u> to decrease your fat tissue and increase your lean tissue (muscles, bones and fluids). You can't do this unless you combine a personalized diet with a personalized exercise plan.

I'm purposely not putting in a height-age-sex chart for "ideal weights." (You've seen these in diet books and physical fitness manuals—they tell you what your "ideal" weight should be.) Your ideal weight isn't a function of just your height, age and sex. Your skeletal build is also an important factor. But more important than your total body weight is the proportion of lean tissue and fat tissue. It's possible, through diet and exercise, to lose a considerable amount of fat without losing any weight, because the increase in lean, muscle mass could balance the fat loss. On the other hand, it's possible, through diet alone, to lose enough weight to put you into the "ideal" range, but still leave you with an unhealthy body compositon that's too high in percent fat.

Your doctor has ways of measuring your percent fat. He can tell you whether you fit into the standard range (16% to 19% body fat). If you're above standard—even if your total body weight is normal—you may run an increased risk not only of heart disease, but also of diabetes, cirrhosis of the liver, hernia, intestinal obstruction, and other health hazards. So don't gauge the state of your health simply by what your bathroom scale tells you. If you follow your doctor's advice for maintaining proper body *composition,* your proper *weight* will follow naturally.

The Benefits. I probably couldn't list all the direct and indirect benefits of a good physical conditioning program based on regular exercise, healthful diet, and adequate rest, but all of the following benefits either are medically recognized or have been reported by police, sheriff, and highway patrol officers participating in conditioning experiments:

- *Improved motor ability.*
- *Improved cardiorespiratory fitness.*
- *Better circulation.*
- *Weight control.*
- *Sense of well-being.*
- *Less fatigue.*
- *Better response to sudden physical and emotional demands.*
- *Better posture.*
- *Fewer lower back problems.*
- *Fewer line-of-duty injuries.*
- *Fewer days on sick time.*
- *More energy.*
- *Improved sleeping and resting.*
- *Feeling better on awakening.*
- *Less gastrointestinal trouble.*
- *Better digestion and elimination.*
- *Less worry about health.*

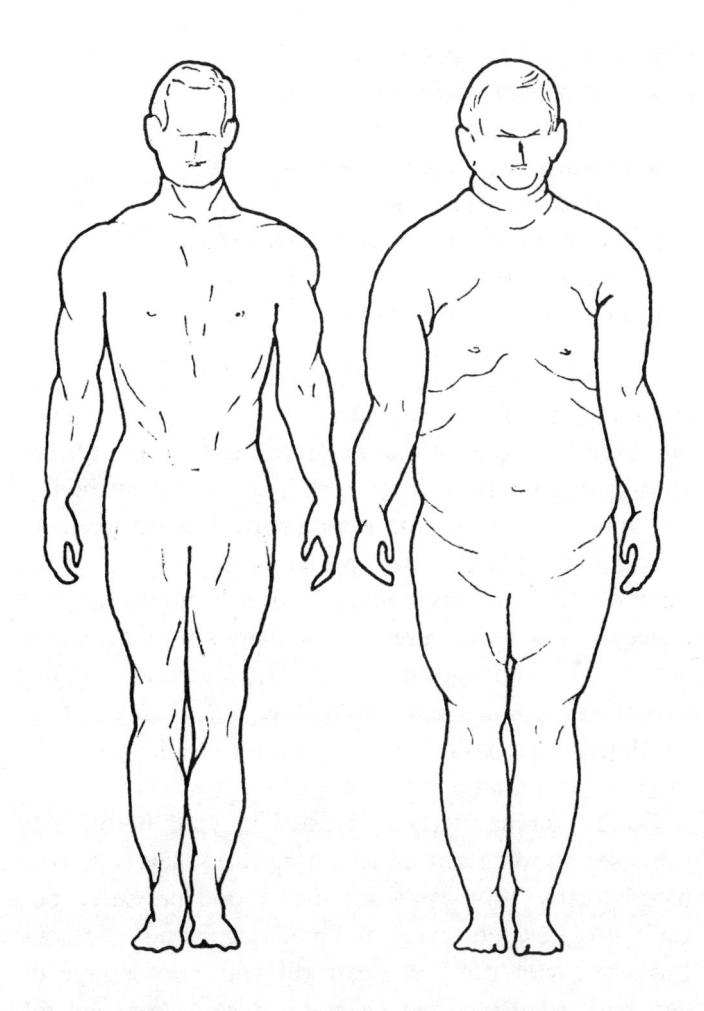

Two people of the same age, height and weight may have completely different body composition, one getting his weight from solid, lean, muscle tissue, and the other having a corresponding weight that's high in flabby fat. Which type are you?

- *Improved self-confidence.*
- *Less tension and easier to relax.*
- *Improved job satisfaction.*
- *Less worry over non-health matters.*
- *Less illness and disease.*
- *Improved ability to survive a heart attack or serious injury.*
- *Improved sex life.*

Physical conditioning is a chore for many people, including me. Paying a doctor for an examination and a diet-exercise prescription, when you don't see anything wrong with yourself, may seem like a waste of money. But there's no legitimate reason for a law enforcement officer (or anybody else) to take better care of his car than he takes of his body. You pull regular maintenance on your car, even when there's nothing wrong with it. You change the oil and filter periodically, just to keep it running right, even though you may not see anything wrong with them. You want to take good care of your car, because you've got a lot invested in it.

Don't you have a lot invested in your body? You can't see the cholesterol and triglyceride levels in your bloodstream. You can't see high blood pressure. You can't see elevated levels of blood sugar and uric acid. But what you don't see can kill you. So splurge on yourself—get a physical check-up once a year, and follow your doctor's advice. You don't have to be one of the 985,000 people who will die of cardiovascular disease during the next 52 weeks. *You can survive cardiovascular risks!* ☐

5

SURVIVING TRAFFIC RISKS

The second biggest controllable risk you face on the job is being killed by an automobile. Although the National Safety Council doesn't keep annual statistics on traffic fatalities by occupation, it's often estimated that more than 200 law enforcement officers are killed each year in traffic-related accidents (that's about double the number who are feloniously slain). Approximately 10 times that number suffer non-fatal (but often disabling or disfiguring) injuries.

It might be natural to expect that law officers would be better-than-average drivers, with lower-than-average traffic accident death rates. That isn't necessarily true, because officers don't follow "average" driving and pedestrian patterns: most average drivers don't spend as much as 8 or 10 hours each day behind the wheel; they aren't constantly getting in and out of a car in traffic; they don't have to stand in busy intersections directing traffic; they aren't regularly out walking around in streets at accident or disaster scenes; and they don't have to engage in emergency and pursuit driving. The average person doesn't have *your* level of exposure to traffic risks, so national averages don't readily apply to you.

Just as it would be a mistake for an outside observer to assume that law enforcement officers are immune to traffic risks, it would be a mistake for you to make that erroneous assumption about yourself. You may be above-average in driving skills, but you're also well above-average in traffic risks due to the nature of your work.

So even though you're the one who watches everyone else's driving and investigates everyone else's accidents, you've got to be especially careful about your *own* driving. To improve your survival odds against the traffic accident risks which face law enforcement officers, consider the following.

ROUTINE PATROL DRIVING

Many officers and supervisors automatically assume that the greatest traffic accident danger comes from pursuit driving or Code 3 runs. Actually, a survey by IACP found that 90% of all officer-involved accidents occur when the unit is either parked or on normal patrol, with the remaining 10% due to emergency

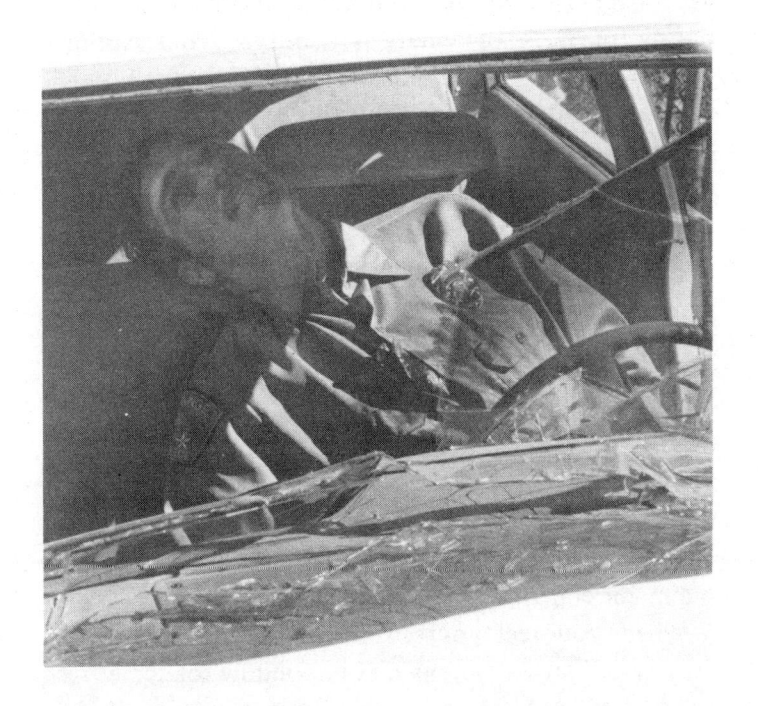

Traffic accidents kill twice as many officers as armed criminals do. The risk of death in a traffic-related accident is the second largest controllable threat faced by law enforcement officers.

driving. An agency in a large Midwestern city examined officer accident records for one year and found that 78% happened during routine patrol duty and 22% during pursuits or emergency duty. And National Safety Council figures show that 50% of all traffic injuries result from accidents involving impact speeds of less than 40 MPH. So don't underestimate the dangers of low-speed driving. In addition to observing the rules of the road that you enforce on others, watch out for your special occupational risks.

Fatigue. If you aren't getting 7 hours of sleep each night (or day), and if you aren't exercising regularly, you will become fatigued from lack of rest and muscle inactivity. Fatigue can slow down your reflexes, thereby increasing your reaction time, and it can diminish your visual efficiency, thereby interfering with your ability to promptly perceive road dangers. Fatigue is a major contributing cause of patrol accidents.

When you feel yourself slowing down and getting drowsy or inattentive, open your window for some fresh air, and get out of the car every few minutes and walk around (shake doors, or chat briefly with merchants). If you have a partner, talk to him; if you don't, sing out loud to yourself or recite something. Talking aloud will perk you up. When you go off duty, get some sleep, and try to put in an exercise session before your next shift (although you might think physical exercise would add to your fatigue, the opposite is true—it stimulates your circulation and improves muscle tone, countering the effects of fatigue).

Distraction. If your mind is still on the last call you handled, instead of on your driving, or if you're preoccupied with worries or fears or other distracting emotions, you won't be paying proper attention to what's coming up the road at you. The same thing holds true if you're trying to talk to your partner with hand gestures, or turning to look at his facial reactions as you tell your story.

If something is weighing heavy on your mind, stop your car at the side of the road and think it over for a minute or two, reach some tentative decision, form a plan to think about it some more on your coffee stop or lunch break or end of watch, and then put your mind back onto the business at hand.

Learn to talk to your partner with your *voice,* and without hand gestures or head-turning movements. You should be coordinated enough to carry on a conversation without taking your eyes off the areas you need to constantly survey for impending problems.

Double-Duty Looking. Unlike the civilian motorist who only has to watch the road as he drives along, you also have to be alert to things that are going on *off* the road. In addition to watching for traffic hazards, you have to be scanning the neighborhood off to your left and right for criminal activity; you have to give pedestrians a quick once-over; you have to notice license plates, vehicle descriptions and occupant descriptions for mental comparisons with your hot sheet; and you have to be watching other traffic for equipment and driving violations. So your eyes have to do double duty, doing both driver-looking and police-looking.

The 2 ways for you to accommodate this double duty are to *drive slower* and *look faster.* If 35 MPH would be a safe speed for a civilian motorist, reduce your speed to 30 or 25. This will give you more time to do your police-looking over a given distance. And learn to move your eyes without turning your head to survey the area as you drive through it. This will let you look faster (you can move your eyes from side to side faster than you can turn your head from side to side).

If something requires closer attention, don't fix your gaze on it as you drive along—pull over and stop to look at it. Even at 30 MPH, your 2 tons of steel will be rolling ahead at the rate of 45 feet per second. If your eyes are off the road for 5 seconds, your car will have traveled 225 feet with no one in effective control. That's just asking for an accident to happen.

Right of Way. Most drivers, as you know, drive extra carefully when they see your marked car in the vicinity. Many of them go so far as to give you plenty of room and yield their right of way to you, even when you're not on a priority run. Officers who get accustomed to this sort of special treatment may come to *expect* it from all drivers. That becomes a danger the day a driver who isn't impressed with your marked car demands his legal right of way and collides with your expectations.

To avoid this danger, don't get into the habit of thinking you're entitled to special right of way consideration when you're on normal patrol. In fact, you should even be willing to surrender your own legal right of way when it's safer to do that than to insist on it and create an accident risk. After you've let someone violate your right of way in the interest of safety, you can always pull in behind him and write him for the violation.

Drive Ahead. Depending on your speed, and whether you're patrolling a city street or a highway, you should drive half a block to a mile ahead. You need to know what's happening down the road, because you may have to start taking some response to upcoming conditions long before you get there. In addition to watching to your sides and rear, you should be looking well ahead for stalled vehicles, brake lights coming on, vehicles entering from cross roads, vehicles pulling away from curbs (watch for wheels turned into traffic, exhaust fumes and driver movement), road debris, narrowing lanes, parked cars, pedestrians, bicyclists, bumps, dips, potholes, oil slicks, water puddles and ice patches. Don't focus all your attention on the car directly in front of you (that driver's probably so glued to your image in his rear view mirror that *he* isn't aware of looming hazards, either).

Weather Conditions. Light rain on a dirty, oily highway will create a slippery surface. Reduce your speed, and start stopping well ahead. As with roads covered with ice or snow, pump your brakes with quick, short stabs, rather than applying continuous pressure.

If the pavement is covered with water (rain, melted snow, broken water main or fire hydrant), your tires may ride on a layer of water, rather than on the pavement. This effect is called "hydroplaning," and it can begin to occur at speeds as low as 35 MPH. At a speed of 55 MPH, your tires may be completely out of contact with the pavement, which leaves you completely out of steering control. You have the same problem intermittently when you drive down an otherwise-dry street that has standing puddles, over which your car will hydroplane.

To keep from turning your car into a surfboard, make sure you have deep tire treads, and *reduce your speed* before driving onto wet pavement.

Ice patches stay longer in shaded spots, on top of bridges and overpasses, and in tunnels and underpasses. Reduce your speed before you get to these places during winter driving.

Blind Spots. The full field of vision you normally have depends on the overlapping fields of both eyes. When the field of one eye is partially obscured (such as by the corner post supporting your windhshield), the overlap is incomplete, and you have a blind spot. At a particular angle, you won't be able to see a vehicle coming from your left or right. To compensate, pause just a moment longer at stops. As the vehicle moves across to your left or right slightly, it will pass your blind spot and appear in your field of vision.

You also have a blind spot in your right rear on a multi-lane street or highway, even if your car has a right side mirror. Don't depend on mirrors (either side) to tell you when it's safe to turn or change lanes. Always glance over your shoulder before moving your car on the roadway.

EMERGENCY AND PURSUIT DRIVING

According to the National Safety Council, approximately 250,000 pursuits each year produce nearly 8000 crashes, 5000 injuries, and more than 500 deaths of civilians and officers. Other forms of emergency driving also produce deadly statistics. Although 50% of injuries occur in low-speed accidents, an accident at the higher emergency speeds is more likely to produce a more serious injury, or death.

I can't teach you pursuit driving in a book. No one can. You need behind-the-wheel instruction and practice, under the guidance of experienced, high-performance drivers. I strongly suggest that if you haven't already done so, you look into the possibility of attending a police driving academy where you can get this practical instruction.

What I *can* do for you here is to describe the common dangers and precautions of emergency driving.

Time, Speed and Distance. Basic to your understanding of risk evaluation on the road is an appreciation for how little time you may have to react to a risk, and how much ground you may be covering before you can complete your risk-avoidance technique. I don't know how they measured it, but traffic experts say that your *perception* time, on the average, is ¾ of a second. During perception, your eyes see images and transmit them to the brain via the optic nerve, and the brain perceives these images as a danger, requiring you to take some sort of evasive action.

The experts have also measured your *reaction* time as ¾ of a second. During reaction, your brain evaluates your alternatives (Should you brake? Speed up? Turn? Honk? Do nothing?), decides on a course of action, and sends impulses to your muscles to take the selected action.

Before you ever put a risk-avoidance plan into action, therefore, your car or motorcycle has continued in the same direction, at the same speed, for 1.5 seconds. If your speed was 35 MPH, you've come 78 feet closer to the risk before you can do anything about it; if your speed was 50 MPH, you've come 110 feet closer; if your speed was 70 MPH, you've come 154 feet closer to danger. (To convert MPH to feet per second, multiply by 1.5: 10 MPH = 15 feet per second, etc. To compute perception or reaction distance, multiply MPH by 1.1: 50 MPH = 55 feet in ¾ second perception or reaction time.)

If the evasive action you decide on is braking, you need a certain amount of distance for the skidding to bring you to a halt. Add the braking distance to the perception and reaction distance, and you have the total stopping distance. Here are 3 examples:

MPH	Perception +	Reaction +	Skid =	Total*
35	39'	39'	58'	136'
50	55'	55'	119'	229'
70	77'	77'	233'	387'

*Assuming clean, dry pavement, with 0.7 coefficient of friction.

Now you can see why it's so important to leave a lot of space between you and the car ahead, and to watch well ahead for possible hazards in your path.

Code 3 Rules. When your department regulations authorize your use of red or blue lights and siren on an emergency call, you will be driving at an increased level of risk. Other drivers are not practiced at accommodating emergency vehicles—they aren't accustomed to your busting stop signs and red lights, driving on the wrong side of the road, and coming out of nowhere at high speeds. Some of them get so flustered that they forget to pull over to their right. Some drivers are so preoccupied inside a noise-filled dream world that they won't see or hear you at all. Still others will notice you at the last minute and take sudden, unpredictable actions to get out of the way.

Unless you *always* pass to the left of traffic going in your direction, you'll confuse drivers as to what they should do.

Unless you *always* slow way down or even stop momentarily and yield to intersection traffic when you hit a red signal, you run the risk of being broadsided by a driver whose radio was louder than your siren.

Approaching a green light, you should take your foot off the accelerator and have it poised above the brake pedal, prepared to stop if necessary. Once you get into the intersection, *get out quickly.* Your speed should be slowest just *before* entering—not while crossing—an intersection.

If your siren is manually operated by the horn ring

(not electronic), use the full range; some people may not hear either the lower or higher frequencies.

Dispatchers should avoid sending multiple units on Code 3 runs, especially if their respective locations put them at right angles to each other as they approach the scene. Two such units assigned to an "officer down-shots fired" call in Phoenix once turned a single-fatality call into a triple-fatality situation when units at right angles, their sirens cancelling each other out, smashed into each other and killed 2 responding officers.

Be prepared for the unexpected. Don't assume that everyone is going to get out of your way, just because you've hit the light bar and siren. When you're driving at increased *risk* levels, drive at increased *alertness* levels.

Pursuit Principles. The first rule on pursuits is to try to avoid them. If you're planning a felony stop, or anytime you can reasonably predict that an attempt to stop a car might result in a pursuit, try to time the stop in an area where you give the driver the least opportunity to jackrabbit. The cul-de-sac or dead-end street, of course, is ideal. A straight street with natural or man-made barriers along the sides (hills, ditches, high fences, walls) is better than one with many connecting side roads, alleys and private driveways. (For tactical reasons, you won't want to give the driver a chance to split *on foot,* either, so avoid spots adjoining woods, high grasses or crops, or commercial or residential "mazes" into which he can flee.)

You can't always find a good spot, and you can't always predict who's going to rabbit. The start of a chase is usually sudden and unexpected, so you should be prepared *ahead of time.* Always drive with your seat belt fastened and your helmet on your head. Be sure your seat is properly positioned so that your hands and feet have good access to the controls. Don't have loose objects lying around on the seat or the floorboard—a sharp turn may put them in your lap or between your feet, and a crash may put your ticket book into your eye or your briefcase into your nose. All these items should be carried in a secured restraining bucket or be strapped down.

Don't start a pursuit when your fuel gauge says "empty." If you lose power just as you're trying to corner at 50 MPH, your power brakes and steering won't be there to keep you from rolling.

During the pursuit, keep *both* hands on the wheel as much as possible, at the 3 and 9 o'clock, or 2 and 10 o'clock positions. Stick your microphone inside your collar, against your neck, and give the dispatcher a description of the pursued vehicle, the license number if you're close enough to read it, and the location, direction and speed of the pursuit. Don't stay on the air all the time, or the dispatcher can't communicate with other units for assistance. Give interval reports about every mile or so.

Don't tailgate the pursued car. Stay a safe distance behind and wait for the driver to give up, or run into a roadblock, or crash and bang himself up. When he loses control and spins out, you don't want to be right on top of him.

The most dangerous high-speed maneuver is turning

from a straight path, because centrifugal force acts on your car to keep it going in a straight line. The rule for curves and corners is to "go in slow and come out fast." Reduce speed before you begin to turn, then accelerate out, as the curve or corner permits, to overcome the centrifugal force (you need to practice this maneuver at the driving academy).

If you take a curve too fast, you can go into a power skid and fishtail. When that happens, ease off the accelerator and let the car stabilize, then countersteer (steer in the same direction you're skidding) until you stop or straighten out.

Don't cut corners, and don't drive too close to the edge of the roadway near soft shoulders. If you *do* run off the road, try to continue in a straight line as you decelerate, and then gradually correct back onto the pavement. Don't make a sudden, jerky turn the instant you run off the road, or you may lose control.

If possible, come to a stop gradually, using pumping action on the brakes. If you jam on the brakes too hard and go into a locked wheel skid, get off the brakes. If your brakes are out of adjustment, either the front brakes or the rear brakes may take hold before the others and put you into a braking skid.

In a rear wheel braking skid, your car will spin 180 degrees around, and you'll be going backwards. Release your brakes and countersteer.

In a front wheel braking skid, your car will continue straight ahead, no matter which way you steer. Get off the brakes. Let up on the accelerator, and steer in the same direction as the skid. (All of these emergency stopping, turning, and skid control techniques should be practiced on a driving course.)

If you lose your brakes and the way is clear, just coast to a stop. If you have to get the car stopped faster, pull your handbrake, downshift, and if necessary, sideswipe unoccupied parked cars, embankments, or substantial buildings.

If you have a blowout, hold the wheel tightly and take your foot off the gas pedal. Don't jam on the brakes. Hold the car in line and slow down gradually.

If you crash, pull your feet back against the car seat, grip the steering wheel tightly, and let your seat belt hold you in place until the car stops moving. Then get out quickly and move at least 50 yards away from the car, in case of fire or explosion.

If you are submersed, unfasten your seat belt and keep your windows rolled up. If you roll a window down too soon, the rush of water into the passenger compartment will be too strong to allow you to escape. Keep your head high and wait until water gets up to your chin. Then take a deep breath, crank your window down (the in-rush won't be very strong now), and swim out and up.

VEHICLE SAFETY

To give yourself the best possible chance of avoiding or surviving traffic accidents, be sure you start the shift with a safe vehicle. Don't take somebody's word for it that the car was alright on the last shift—before you put *your* body inside, give the car *your* safety check:

- ☐ Check the brake pedal action.
- ☐ Test lights, horn, and emergency equipment.
- ☐ Check windshield wipers.
- ☐ Make sure windows and mirrors are clean.
- ☐ Check tires for tread depth and inflation (low pressure weakens the sidewalls and can increase the chance of a blowout).
- ☐ Check the steering action.
- ☐ Make sure the seat stays where you put it.

SAFETY DEVICES

Three pieces of equipment can help save your life in a crash. The first is a helmet. If your department provides you with a helmet, don't leave it lying on the seat. It can't protect your skull unless it's on your head.

Another item is body armor. You may only have thought of body armor as protection against a gunshot or knife wound, but it has also been known to save dozens of officers from death or serious injury in car crashes. Your vital organs don't know the difference between being stabbed by an assailant's knife and being sliced up by a piece of broken windhsield or twisted steel. Body armor doesn't know the difference, either.

The third lifesaver is your seat belt. The objections to using a seat belt are so standard that they've been catalogued by the NSC; these objections are so misguided that they're easily rebutted by facts and common sense about survival:

Objection: *"Most of my driving is at low speeds—I don't need a seat belt for that."*

The Fact:	Half of all injuries occur at speeds below 40 MPH.
Objection:	*"Being thrown clear may save my life."*
The Fact:	You chances of survival are 5 times better if you stay in your seat.
Objection:	*"I could be caught strapped in during a fire, or if I go underwater."*
The Fact:	Of all injury accidents, only 0.2% involve fires, and only 0.3% involve submersion. Your chance of escaping these dangers is enhanced if you are not knocked unconscious by being slammed around inside the car.
Objection:	*"I get in and out of my patrol car 50 times a day. I can't be constantly buckling and unbuckling my seat belt. That's just a time-consuming nuisance."*
The Fact:	It takes 2 seconds to buckle your belt and ½ second to unbuckle it. That's very little time to spend for the added protection you get. Once it becomes an automatic habit, you won't even notice it.
Objection:	*"I'm a good driver. I've never had an accident in my life."*
The Fact:	Neither had 80% of those who become involved in the 18 million traffic mishaps

that kill 52,000 Americans every year.

Objection: *"Seat belts don't do any good, anyway."*
The Fact: *NSC experiments and studies show that use of seat belts would save 12,000 lives this year. One of them could be yours.*

Seat belts, body armor and crash helmets don't do you any good unless you use them. And you don't often get advance warning that you're about to need them, so you have to develop the personal survival habit of being prepared all the time by wearing your lifesavers.

PEDESTRIAN HAZARDS

Outside your car, without a steel barrier surrounding and protecting you, you're even more vulnerable to injury or death if you're involved in a traffic accident.

Traffic Stop. When you make a traffic stop on a city street, try to engineer it to a place where the road is wide, where there are no blind curves or hills to your front or rear, and where you'll be able to step out of the road to talk to the driver and write your citation. You want to interfere with normal traffic flow as little as possible, and you want approaching traffic to be able to see you from a distance, so they aren't forced into last-minute lane changes to get around you.

Park your car to the left rear of the violator's car, giving yourself a 3' offset safety corridor for your walk-up approach. Set your parking brake before getting out of your car.

Don't open your door on the traffic side until it's safe to get out *and* to get to the front of your car, into your safety corridor. Don't leave your car door standing open on the traffic side.

Don't stand out in the highway beyond your safety corridor as you talk to the driver at his window. If he starts to get out, don't back out into the traffic lane.

Don't stand to the traffic side of your car, writing the citation on your hood. Don't stand between your car and the violator's car—this is the "leg smash and amputation zone" if someone plows into the rear of your car and shoves it forward.

When you're returning to your car to use the radio, or after issuing the citation, check again to see that you have time to get inside and close your door before oncoming traffic reaches your location.

If you're making a traffic stop on a narrow highway or high-speed freeway, pick a location that allows both you and the violator to pull completely off the right side of the road. Do not use the left-hand offset unless the shoulder is wide enough to permit it without obstructing the traffic lanes. If traffic is heavy and close, get out through the front passenger door and have the violator leave his car the same way.

Before re-entering traffic after the stop, check to be sure you have sufficient clearance.

(In addition to these traffic safety considerations, there are *tactical* considerations of the carstop location, approach and techniques. These are covered in Chapter 7.)

Accident Investigation. When you're out of the car investigating a traffic accident, don't expose yourself to being run over in the middle of the road. If the involved vehicles can be moved out of the roadway, clear the road and conduct your investigation on the shoulder or in a parking lot. If the road can't be cleared quickly, set out warnings to approaching traffic to get them to slow down: use your unit's emergency flashers, set out reflective warning devices, or set up a flare pattern, if the situation permits *(do not ignite flares if there is gasoline on the roadway from the wrecked vehicles)*.

Stay out of the road during your interviews with drivers and witnesses. Be sure you have adequate clearance in both directions before getting out in the road to do your measuring, photography, debris collection, etc. Have your partner or back-up officer, if available, watch traffic for you and warn you when to withdraw.

Don't get in and out of your car on the traffic side unless it's safe to do so.

Directing Traffic. Standing in the middle of a busy intersection with a whistle and a pair of white gloves isn't very interesting duty, but it *is* dangerous. For your own protection, keep the traffic speed down with your whistle and hand-and-arm signals. The faster you let them whiz by you, the more dangerous your situation is.

Be sure you're not off-center. Any position other than the very center point of the street or intersection will put you closer to the traffic from one direction.

If visibility is a problem (for example, during fog or bright sun glare), consider wearing a bright orange or yellow plastic vest so motorists can see you.

* * *

Whether inside your car or outside on foot, don't assume that people will see you and avoid endangering you just because you're in uniform and use a marked car. Don't get to feeling that you're invulnerable. You're not.

Instead, take the attitude that you're going to handle your heightened levels of traffic accident risks with heightened levels of awareness, preparedness, proficiency and precaution. Learn to control your vehicle, anticipate danger, and don't expose yourself to pedestrian injury. *You can survive* traffic risks! ☐

NOTES ON LOCAL RULES

6

SURVIVING SUICIDE RISKS

Suicide is the 9th leading cause of death in the United States, accounting for at least 35,000 deaths annually (because many suicides are covered up and listed as accidental or natural deaths, some experts believe the true figure is closer to 100,000 suicides per year). Law enforcement officers seem to be at higher risk than the general population: one study of suicides within the New York City Police Department found officers killing themselves at a rate 6½ times higher than the national average, and the last occupational suicide study by the National Center for Health Statistics found the law enforcement suicide rate nationally to be 83% higher than for the general population. By the most conservative estimates, the law enforcement officer is at least 10% more likely to be killed by his own hand than by a criminal.

Unlike traffic accidents, cardiovascular diseases, and homicides—the 3 causes of death which I've called *largely* preventable—suicide is *completely* preventable. It's the only 1 of those 4 controllable risks which you and you alone have the absolute power to prevent. In a sense, that makes the suicide death the most tragic of all. It should be the one risk that's easiest for you to survive, and yet the statistics say that it isn't. The statistics show that you're significantly less likely to survive this risk than the risk of a fatal field encounter. Such a tragic state of affairs should be disgustingly unacceptable to you.

So if you were just about to say to yourself: "This chapter doesn't apply to me . . . I'll skip over to the next chapter," don't do it. You can read this chapter in 15 minutes. (Before you finish the chapter, 10 people in the United States will attempt suicide; 1 of them will succeed. By this time next week, 2 officers will have killed themselves.) This chapter may not apply to you today, but before your police career is over, something in here may help you save your own life, or that of a fellow officer. That possibility is worth an extra 15-minute investment.

WHO COMMITS SUICIDE?

All kinds of people, of both sexes, all races, and all ages commit suicide, but not all at the same rates. Although more women attempt suicide, more men succeed. More whites than blacks kill themselves, more city dwellers than urban residents, more single and divorced people than married ones, more people without children than those with, more Protestants than Catholics, and more Catholics than Jews.

A study of 93 police officer suicide cases showed the following statistics:

- *64% of the officers were patrolmen...*
- *83% were in the age group of 30 to 45 years...*
- *75% were married...*
- *28% were qualified to retire on pension...*
- *38% had suffered a recent departmental problem...*
- *51% had suffered a recent marital problem...*
- *35% were "aggressive, impulsive" officers...*
- *65% were "quiet, reliable, good cops..."*
- *90% used their service weapons to kill themselves.*

WHY DO OFFICERS KILL THEMSELVES?

"I just can't believe it. He had everything to live for. Why in the world would he do it?"

That's a typical reaction for those who unexpectedly learn that a fellow officer has committed suicide. They can't understand what made him do it (only about 1 in 7 suicides leaves a note explaining his reasons). But the reason isn't usually too mysterious: some officers commit suicide as a means of revenge against family members who will suffer from the death, and other officers kill themselves because there's something about their continued life that they just don't want to keep living with.

The occupational and marital stress factors we discussed in Chapter 3, if they are allowed to take their toll, can build up to the point where they begin to seem unbearable. If a "final straw" is added to this burden, it can send a seemingly-stable officer into fairly sudden personality disintegration and states of depression, from which suicide begins to appear an attractive way to put an end to trouble.

Examples of such "final straws" that regularly appear in suicide cases are demotion, departmental discipline, fights with the spouse, separation, divorce, loss of a loved one, and learning of a serious or terminal illness. Contributing factors, according to most experts, include the increased mobility and lack of roots of Americans in general, the deterioration of the family structure and the changing roles of men and women, the self-isolating emphasis on independence, the dissolution of social institutions, and the exaggerated insistence upon job promotions and financial success.

One strikingly recurrent situation among those 93 officer suicides studied by Dr. Paul Friedman involved a "nagging wife" problem. In one case, a 30-year-old patrolman, whose father had committed suicide as a patrolman, was being nagged by his wife about a trivial matter. The patrolman said: "OK, I'll get out of your way," and promptly shot himself.

Another case involved a 32-year-old patrolman who shot and killed himself in the presence of his 11-year-old daughter and his complaining wife; one year later, the daughter committed suicide by leaping from a window.

In still another case, an officer was being scolded by his wife for buying a lavish restaurant meal for visiting relatives, when he pulled his revolver and blew his brains out. Other married officers killed themselves after hearing their wives complain about their debts, their low standard of living, the husbands' lack of promotion, and friends and relatives who were financially better off. One officer shot himself in a phone booth, while arguing with his wife over the phone; his last words were: "Listen to this!"

These are the kinds of cases in which officers seek to inflict pain, suffering, humiliation, and feelings of guilt on their surviving spouses. That these officers felt this kind of anticipated revenge was worth the taking of their lives is an indication of how distorted their mental outlooks had become.

Also among the 93 suicides studied were a total of 10 officers who killed themselves because of a *mistaken* belief that their illnesses or diseases were incurable.

THE WARNING SIGNS

Although a suicide may appear at first glance to have been unpredictable and totally unexpected, almost all suicidal officers have left a trail of warning signs that neither they themselves nor their friends and families properly recognized and heeded. You should know what these common danger signals are, so you can watch for them in yourself and in fellow officers.

Attempts and Threats. Those who study suicides report that 80% of all persons who kill themselves have previously given obvious clues of their intentions. That someone would say: "I'm going to kill myself," or "You won't have to worry about me much longer," or "You're going to find out what it's like when I'm dead," is a *serious* indication that he is at least considering suicide. If it isn't *taken* seriously, tragedy can result. Many officers' wives have learned that lesson too late, after their taunts of "Go ahead—you don't have the guts," proved to be the final straws.

An actual *attempt* at suicide is an even stronger warning of a person's potential for self-destruction. A common reaction to the suicide attempt is for an observer to say: "He's just trying to get attention." And while that observation is precisely correct—the attempter desperately wants to dramatize his despair—the fact that he would risk his life in a plea for help shows how very badly he *needs* it.

An *attempt* at suicide is the most obvious danger signal. Unfortunately for law officers and their loved ones, the method usually chosen by officers for their attempts—the firearm—doesn't leave many attempters alive to be helped.

Depression. Mental depression is a condition which affects an estimated 30 million Americans. Its symptoms are often found in the backgrounds of many people who commit suicide. Here are some things to watch for:

- *Sleep disturbances—either insomnia or oversleeping, but especially early awakening without being able to get back to sleep.*
- *Loss of appetite and abnormal weight loss. All food seems to be tasteless, or all tastes the same.*
- *Frequent indigestion or constipation.*
- *Loss of interest in normal sexual activity.*

- *Loss of interest in work, family, and hobbies.*
- *Lack of ability to concentrate.*
- *Lethargy—no drive to get anything done.*
- *Nervousness, anxiety, and sudden crying for no apparent reason.*
- *Tendency to stop communicating with friends, family, and co-workers, and to withdraw into isolation.*

While some of these symptoms are often caused by things other than depression, the simultaneous or progressive occurrence of several of these conditions should be taken as a strong indicator of the development of depression, with suicidal tendencies possible.

Personality and Behavior Changes. People who are contemplating suicide often undergo visible personality changes. For example, the 65% of officer suicides who fit into the "quiet, reliable, good cop" category had always been thought of as "one of the gang" by their fellow officers, and had appeared to supervisors to be passive, conscientious officers, with everything under control. But in the final days before their suicides, these officers had become irritable, suspicious and moody—so much so that their co-workers realized something was very wrong.

The 35% who were described as "aggressive and impulsive" became increasingly reckless, often drinking excessively, taking unnecessary risks, and becoming sexually promiscuous. One of the officers in this group became so deranged that he shot and killed his partner for refusing to drink with him, and then he turned his gun on himself.

Most of the officers from both groups had shown increased use of alcohol, more frequent work absences on "sick days," and an increased record of stomach ailments. Although they had not been diagnosed as suicidal, 67% had established some record of disturbed mental status.

Final Arrangements. Revealing clues that a person has come to a decision to take his life are evident when he starts making arrangements for his death and the distribution of his property. Watch for these signs:

- *Sudden interest in making out a will.*
- *Revision of life insurance coverage and assembling policies together.*
- *Purchase of a burial plot.*
- *Acquiring the means of death (buying poison, barbiturates, hose for auto exhaust, new bullets, single-edge razor blades). This step is usually unnecessary for law officers, who routinely carry their means of death.*
- *Writing letters and making "farewell" phone calls to old friends and faraway relatives.*
- *Giving away valued personal possessions, such as cars, motorcycles, skis, cameras and gun collections.*
- *Making unusual cash gifts to friends, family and charities.*
- *Converting assets to cash to pay off mortgages.*
- *Setting up trust funds for children.*

Again, although some of these acts may simply be the result of prudent estate planning, their sudden, simultaneous or progressive occurrence should be viewed as danger signals.

SUICIDE PREVENTION

When you begin to notice the danger signals in your own conduct, or in a fellow officer, what should you do? The first rule of prevention is this: DO SOMETHING! The status quo is what's driving you to consider suicide; therefore, don't continue with the status quo—CHANGE SOMETHING!

Try to identify the problem. Is it job dissatisfaction? If so, talk it over with co-workers and supervisors. If you can't talk to them, or if they're helpless to make changes, go to the top. I don't personally know every chief, sheriff, and commissioner in this country, but I know what kind of men and women they are—not one of them is too busy to talk to a member of his command about departmental problems which may be creating intolerable stresses for field officers. It doesn't matter whether you're in a 2-man department or a 35,000-man department, you don't have to kill yourself to call attention to a problem.

Is the trouble with your spouse? Talk it over. Let your wife or husband know how very serious the situation has become. If you haven't tried yet, go see a marriage and family counselor. On the other hand, if you've given a bad marriage your best efforts and it isn't doing what a marriage is supposed to do, admit your mistake, accept the inevitable, and part company as amicably and as promptly as possible. If you think there's some hope, consider a trial separation. Just don't go on as usual. If the usual situation is tempting you toward suicide, you have to make some kind of *change*.

Your own ego is one of your greatest sources of strength. Use it. Recognize your own value. Look at your own accomplishments. View yourself as a worthwhile person—as a special person—because that's what you are.

Friends, family members and co-workers who are not part of the problem are good resources to help you find the solution. Use them. If you let them know that you need their help, they'll be glad to give it.

If you have a religious faith, it can be a very strong source of support for you. Use it. Talk over your problems with your clergyman. He will be understanding and encouraging.

There are some 200 suicide prevention centers in the United States, with an equal number of "hot lines" staffed by trained counselors. Use them. Visit or call as soon as you realize that you or a friend show the warning signs.

Family doctors, mental health clinics, crisis intervention centers, clinical psychologists, psychiatrists and psychiatric social workers are all sources of help. All you have to do is ask for it. If you suspect depression, see a psychiatrist or neurologist. Depression is often caused by a chemical imbalance in your nervous system, which can be medically treated.

If I were a person facing a recognized suicide risk and exhibiting some of the symptoms, I would not limit myself to any single source of assistance, but would

When you see the warning signs of suicide, TAKE ACTION. Get help. Don't let the situation get worse before you take steps to correct it.

seek help from *all available sources*. An untrained or insensitive person to whom you look for support may take the wrong approach and say the wrong thing, thereby intensifying your feelings of guilt or failure or hopelessness. Even some physicians and psychiatrists may fail—a full 75% of those who commit suicide have seen a doctor within 4 months of the day they take their lives, and the well-known fact that psychiatrists have the single highest suicide rate of all occupations does not particularly recommend them as experts on suicide prevention. My layman's suggestion is that you not place total dependence on anyone, but seek help from everyone. That multiplies the odds of your finding effective assistance.

You should also be aware of the importance of follow-through. Persons recovering from the effects of severe depression sometimes find that they have regained sufficient strength and will to carry out their suicide decisions. Just when friends and family think the crisis is over and the individual is getting to be "his old self," he suddenly commits suicide. One Los Angeles study disclosed that 50% of suicidal patients who were discharged from the hospital killed themselves within 90 days. Continuing, long-range support from friends and family is necessary to overcome the risk of recurrence.

Law enforcement has always been a "macho" profession. There has always been a prevalent attitude that cops stand on their own feet—that they don't need and don't ask for help from anyone. But that's not really a he-man attitude—it's a juvenile attitude. There's nothing grownup about a person who would rather risk self-destruction than ask for help when he needs it. And there's nothing valiant about lying beneath a tombstone that says: "This guy was as stupid as they come, but he sure was macho."

Law enforcement has also been characterized by a closed and close comradeship, in which officers protect one another. But it's sometimes necessary to make difficult decisions about what is and what is not in your fellow officer's best interest. In almost all of the 93 officer suicide cases which I referred to earlier, fellow officers had seen the personality disintegration taking place; they had seen the abuses of alcohol; they had known of the increased extramarital promiscuity; and they had witnessed overreactions and even brutality against citizens.

But in their misguided attempts to be loyal to the mutual-protection code, the officers who had seen all this erratic behavior had *covered it up* from the supervisors and the families of the troubled officers. Friends who could have done something to save a fellow officer's life *failed,* because they were too busy trying to save his *job.* If you care enough to be willing to cover up someone's danger signs, *care enough not to.*

* * *

To reduce the risks of suicide, practice the techniques we discussed to cope with or eliminate stress; keep in good physical condition; practice good nutritional habits; get enough rest; watch for and respond to the warning signs in yourself and others; make changes in depressing situations; and don't hesitate for one second to get help when you need it. *You can survive* suicide risks! ☐

INFORMATION IN THIS CHAPTER IS INTENDED FOR LAW ENFORCEMENT PERSONNEL ONLY. DO NOT PERMIT PUBLIC ACCESS. DO NOT COPY.

7

SURVIVING FIELD THREATS

There are no reliable statistics on the number of murderous assailants who are killed each year by officers who either receive no wounds, or survive their wounds. Statistics *are* reported, however, on the results of those encounters which end in the officer's death. In a recent year, the FBI reported 93 officers feloniously killed in the US. *Only 1/3 of those victim-officers got off a shot at their assailants, and only 3 of the 93 officers managed to deliver fatal fire at the offenders.* In other words, in 90 out of those 93 times, the criminal won the fight.

Why did law enforcement officers respond so ineffectively, or not at all, in 90 out of 93 deadly attacks on themselves? In fact, why would a trained, armed officer *ever* lose a life-and-death battle with a criminal opponent? All of the answers come under a single heading: *inferiority*.

SURVIVAL OF THE FITTEST

It was the English philosopher Herbert Spencer (not Charles Darwin) who coined the phrase "survival of the fittest" to describe the results of the struggle for survival among all living things. The rule that the strong survive and the weak perish during competition applies not only in the Asian and African jungles, but also in the asphalt jungles where you and your fellow officers stand guard over civilization. When you are superior to your adversaries, you will survive. When you are inferior, you will perish.

Inferiority takes a number of forms. Those officers who lose the struggle to cop killers may be inferior in physical conditioning, motivation, preparation, or combat resources (including weaponry, firepower, and manpower). When an officer is inferior in more than one of these categories, the odds against his survival are multiplied. Although pinpointing the exact area of inferiority may sometimes be difficult in reviewing officer fatality cases, this much is indisputable: if the officer's survival fitness had been superior to his opponent's, the officer would have survived.

We've already discussed the fact that comparison studies have shown law officers to be in inferior physical shape to prison inmates, and that smaller, weaker men and women are entering the law enforcement profession. If an officer is inferior to his opponent in physical strength, agility, flexibility or endurance, he begins combat at a disadvantage, which can only be offset with *effective* superiority in some other area, such as motivation, preparation or resources. (One of the 93 officers killed that year was choked and beaten to death by an unarmed opponent.)

In this chapter, I offer my suggestions on ways for you to develop superior survival fitness through motivation, preparation, and effective use of your resources. The key to survival, however, remains the sound mental and physical fitness discussed in earlier chapters.

YOU BET YOUR LIFE

If you've read any of my other law enforcement training manuals, you know that I don't waste your time with forewords and prefaces, and that I don't stick a picture of myself at the back of the book, with a lot of vital statistics about my college degrees and "accomplishments" to impress you. I let my writing speak for itself. It either makes sense to you, or it doesn't, regardless of who I am or what I've done.

But before you go staking your life on some tactic that I recommend to you, I think you have a right to question my qualifications. That's what *I* would do if I were in your shoes.

There are hundreds of law enforcement books on the market, on topics ranging from traffic accident investigation to juvenile law to community relations to forensic pathology. Many of these books are written by experienced officers who know what they're talking about; others are written by educators, academicians and journalists who become criminal justice "experts" by interviewing real, live cops and attending lots of seminars.

Books from people who get their information some way other than through personal experience may be fine in some areas. At the worst, their misinformation may cause you to lose a case in court.

But a book that asks you to bet *your life* on its police combat tactics owes you higher degrees of care and accuracy. When you're under hostile fire out in the field, tactics based on classroom theories and seminar speculation won't do. Second or third-hand interpretations won't do. Complicated, time-consuming shooting techniques won't do. Textbook responses that counter your natural reflexes won't do. When your one and only life is at stake, the good intentions of a misinformed survival "expert" simply won't do.

Before I decided on writing this manual, I examined the leading works on officer survival to see whether there was anything another such book could contribute. Not only did I find the disproportionate emphasis on your lowest risk, as I mentioned in Chapter 2, I also found a number of "standard" tactics with which I flatly disagree, from a survival standpoint.

As evidence of their qualifications, the authors of the survival publications I examined detailed their police experience, their academic credentials, their training backgrounds, and their subject-matter "research."

Not one of these authors claimed to have ever fired a shot at a human target. Not one claimed to have ever been in a firefight, a knife attack, or even a fist fight. Not one of them claimed to be a *survivor* of a hostile encounter.

As I said earlier, I think there's something to be learned from any book on survival. I repeat my recommendation that you read every such book you possibly can. But don't take every word you read (including mine) as gospel, to be applied unquestioningly to every situation you encounter. You should evaluate options for yourself and adopt only those tactics which you are capable of applying, and on which you are content to bet your life. Don't allow any writer to place the bet for you.

Now, I said you had a right to question my qualifications—especially since I'm going to disagree with some of the survival techniques you may have been taught. During my police experience, I was involved in only 2 shootings, neither of which directly threatened my life. I did not attend the FBI survival school, nor any state or regional survival seminars. I did not teach classes in officer survival.

My experience at surviving hostile threats came from the same place as many other officers: military combat. I was no Audie Murphy. I was no John Wayne. The first few times I was under fire, I was as scared as a baby rabbit. But from one fight to the next, I gained the kind of survival experience they can't teach at the FBI academy. It's that experience on which I rely in making recommendations to you on tactics.

Obviously, Main Street is not Kontum Province. What works in a military operation doesn't necessarily work for a peace officer whose actions are subject to departmental and court review, who must take the safety of bystanders into consideration, and whose enemy is harder to identify. But once an identified life-threatening situation develops, the differences are not as important as the similarities: whether you're at war with an armed military opponent or an armed robber, a contest between you and your adversary that boils down to killing or being killed involves the same kinds of efforts and the same kinds of emotions.

I don't enjoy telling war stories. In fact, I don't particularly like to remember that I was even there. But before I ask you to bet your life on the combat tactics I recommend, I think it's important for you to know that when I talk about life-or-death efforts or emotions, I'm not theorizing or speculating or repeating something I've read in a book—I'm speaking from the experience of eleven months in almost daily combat.

I've been under fire from just about every kind of conventional weapon, including artillery, mortars, rockets, machine guns, anti-tank weapons, rifles, pistols, grenades, satchel charges, and even misdirected friendly naval and aircraft bombardment. I've been hit so many times I stopped keeping score after the third Purple Heart. I still carry a souvenir in my right leg.

These hostile contacts came in every form, from sporadic sniper fire to suicide-squad attacks to battalion-strength firefights. I've fought in the jungle, in open fields, and in villages. I've exchanged hostile fire at 300 feet and at 3 feet. When all the ammo had been shot up and the grenades all thrown, I've fought hand-to-hand in holes and ditches, where the survivor was the one who beat his opponent to death with a steel helmet or an entrenching tool, or gouged out eyes with a can opener, or ripped throats open with a sharp stick and stuffed in dirt and leaves for suffocation.

Although I'm not ashamed of having killed men and women who were trying to kill me, the only thing I would say I was *glad* of is the fact that *I survived*. Sometimes I survived *because* of my training, and sometimes I survived *in spite of* it. Like every other soldier who survived continuous combat, I learned to differentiate between training tactics that help save you, and those that could get you killed. Things that look logical in pictures and diagrams, and make impressive—even entertaining—classroom demonstrations, are not always possible or practical in the real world of split-second, heart-pounding, puke-swelling, death-dealing actions and reactions. A few times, I've taken cover behind the fallen bodies of men who bet their lives on textbook tactics—and lost their bets.

MOTIVATION

Street survival doesn't begin out on the street. It doesn't begin in a patrol car. It doesn't begin in the station house at the start of each shift. It doesn't begin at the academy. It begins in the mind of the person who makes a decision to apply for a job as a law enforcement officer. That's the point at which a person commits himself to a course that he knows may bring him face-to-face with his killer. That's the point at which the prospective officer accepts the possibility of being required to take human life. It is in the process of deciding to accept these life-and-death risks that street survival really begins.

When you were sworn in as a law enforcement officer, you took an oath that broadly described your duties. Although your oath didn't say so out loud, it carried an understood condition that you impliedly swore to:

> "... and if it should become necessary, to prevent the loss of innocent life or for my own self-defense, I swear that I will kill the threatening individual, rather than to allow him to kill me or others."

If, because of religious, moral or psychological reasons you have *not* made this commitment, your chances of survival in a kill-or-be-killed encounter are exactly *zero*.

There are people out there who won't think twice about killing you to avoid arrest, or to escape. There are people who will kill you for interfering with their domestic arguments. There are people who will kill you to escape a traffic ticket. There are even people who will kill you for no reason at all.

These people don't have any religious, moral or psychological inhibitions about taking your life. They don't have any shooting-policy checklist to go over in their minds before pulling the trigger on you. They don't have anything to justify. They don't have anything to think over. They don't have any mental reservations creating a drag on their reflexes. Five seconds after they decide to kill you, you're going to be dead, in the majority of fatal-encounter cases.

Now, I'm not going to suggest that you become the same kind of cold-blooded, trigger-happy freak as your adversaries—that's neither necessary nor acceptable. And I'm not suggesting that you disregard your department's shooting policy, or controlling statutes. What I do suggest is that you get these considerations, as well as your own personal feelings about killing, into focus once and for all (if you haven't already done so), so that your mind is clear as to when you *can* and *will* take an opponent's life. Settle any doubts or reservations you may have before you hit the street again—once the killing case starts unfolding, there's no time for soul-searching or careful deliberation.

And don't head into a fatal encounter with the notion that you can act civilized and play fair. Killing is not a civilized activity. When you're facing an identified, immediate threat to your survival, everything you've ever learned about fair fights goes out the window. The man or woman who's trying to kill you isn't going to fight fair. So if you're burdened with the notion that you're going to fight like a gentleman, they're going to be burying you like a gentleman. You're not going to be any match for your opponent, unless you're ready to play dirty.

When your very survival is at stake, there's no such thing as community relations or bad press or shooting review boards or the ACLU. You can worry about all those things later, when you're alive and your attacker is dead. You're not going to get the chance to worry about them later if you give them any place in your thoughts while some punk is trying to waste you.

If you're going to survive a threat to your life, you've got to be just as willing to kill your attacker as he is to kill you. You've got to be more determined to succeed. You've got to be ready and able to beat him at his own game. If you're not, you're simply not going to make it. It *is* survival of the fittest.

I labeled this topic "motivation," instead of "will to live," as many publications do. I'm not worried about your will to live. I doubt that any of those ninety-three murdered officers were suicidal. I don't think will to live is a problem. It comes naturally. You don't have to give yourself any pep talk in order to get excited about wanting to live when there's someone coming at you with a hatchet or a blazing .38.

By "motivation," I don't mean you need any motivation to want to stay alive. I mean that you get motivated to shrug off all those *unrealistic* restraints that your civilized life has put in your head, through books, movies and television fantasies, about how you can neutralize your deadly enemies in some lofty, gallant fashion.

All your life, you've been taught to cherish human life. That you qualified to become a law officer proves that you learned to respect civilization's ideals. But while you are sworn and justifiably expected to honor those ideals right up to the brink of death, when your survival is put in jeopardy, you've got to be able to set the ideals aside for a moment—without even thinking about it—and deal rapidly and decisively with the threat to your survival.

The will to live is instinctive. The will to *kill* is not. That's where you need the motivation. Without it, you're going to be inferior to your killer in survival fitness. By the time the FBI reports have you down in the statistics, your killer will be putting in for parole.

PROTECTIVE GEAR

To give yourself the best protection against gun and knife attacks, you should make maximum use of protective items that are available.

Body Armor. Lightweight, concealable, comfortable body armor made of Kevlar can be purchased in both men's and women's styles for about $100. Since its introduction several years ago, it has been credited with saving hundreds of officers from certain death or serious injury in shootings, knifings, and traffic accidents. The increased use of such body armor may be one reason why the number of officer murders has declined by approximately 30 percent. Estimates based on location and type of injury in the 93 officer deaths studied indicates that as many as 46 of them could have survived, if they had been wearing suitable body armor.

Ballistic Helmets. If your department allows helmets as uniform headgear, investigate the possibility of obtaining a ballistic helmet that meets National Institute of Law Enforcement and Criminal Justice standard 0106.00, Type .357 Magnum. This helmet provides protection against a variety of commonly-encountered weapons, including the .38, .45, .357, 9 mm, and 12 gauge shotgun with 00 lead buckshot.

Ballistic Clipboards. Impact-resistant fiberglass clipboards are relatively inexpensive, can be used to support your citation book or report forms, and will provide some protection in absorbing or deflecting small-arms fire, especially from small calibre, low velocity weapons and ammo. Don't rely on a clipboard as a substitute for body armor—it isn't. Use it for an extra measure of protection, along with body armor and a ballistic helmet.

None of these protective devices do you any good if you don't use them. Officers have been killed while their body armor hung in the locker or lay in the trunk. And if you're going to be carrying a gun and getting into action while you're off duty, consider wearing your body armor beneath your civilian clothes—12% of officers killed in the survey year were off duty.

None of the protective devices do you much good if all the bad guys know you're using them. So *don't advertise.* I see newspaper items and TV reports all the time about some local agency proudly demonstrating their newly-acquired ballistic equipment. And I see TV interviews with hospitalized cops who seem anxious to tell the whole criminal world about how their body armor saved the day. If you're going to tell your next opponent that you're wearing body armor, you might as well leave it off, because he's going to shoot you in the head. Don't talk to reporters about your protective devices *or* your tactics—once they become known, they become useless.

WEAPONS AND AMMO

In most departments, service weapons and ammunition are prescribed for you, and you're not permitted to use anything more deadly. If your department allows you to choose your own type of weapon, you want the most effective sidearm you can proficiently handle.

Experiment with reloading devices and manual reloading to see which you're more comfortable with. Then pick one method and stick to it forever—don't alternate from month to month. Speed loaders for revolvers are significantly faster than manual reloading.

Clean your weapon and ammo routinely once a week, and more often as required by the weather and your activity—immediately after firing, or exposure to sandstorms, rain or snow, or when you've been rolling in the dirt with a drunk. (Many survival writers and lecturers feel it's obligatory to include a story about a cop whose gun malfunctions due to inadequate maintenance. I don't. A person who doesn't care enough about his own survival to be sure his gun is operable every time he puts it on isn't going to read this far into this book.)

Select a holster that keeps your weapon in place while you're running and jumping (try it out with an unloaded weapon). Be wary of trick holsters, mechanical assists, "breakfronts," and other unconventional designs that may make it easier for an opponent to draw your gun.

If your car comes equipped with a shotgun, you should regard it as your equalizer. Many suspects are armed with .357 or .44 Magnums, sawed-off shotguns, and automatic rifles—all of which make a typical .38 police special an inferior source of firepower. To improve your survival fitness, be sure you're proficient with the shotgun, and *take it with you* whenever the circumstances indicate that it might be useful. Its appearance alone is enough to tame some would-be adversaries.

Back-up weapons, properly identified and recorded with the department, can also increase your firepower superiority. If you empty your primary gun in a firefight and don't have time to reload, or if your primary gun malfunctions and jams, your back-up weapon can save your life. It also comes in handy, of course, if you lose your primary weapon or become disarmed somehow.

Ideally, both weapons should be the same caliber and type, so that your ammo fits both, straight from your speed loader, and you don't have any mental "switchover" to do. For example, if your primary weapon is a 4-inch .38 S&W revolver, your back-up should be a 2-inch .38 S&W revolver—not a Colt (opposite cylinder rotation), nor a 9 mm automatic (ammo isn't interchangeable, and mental switchover required).

Always carry your back-up in the same concealed place, where it will be secure and accessible. If you use an ankle holster, practice pulling your trouser leg up, dropping onto your opposite knee, and drawing, all simultaneously. Also, practice pulling your leg up

SURVIVING FIELD THREATS/135

Your shotgun can be your equalizer, if you remember to TAKE IT WITH YOU on dangerous calls. Confronted with a view like this one, most suspects will readily submit.

toward your hand while lying on your stomach, back and sides. (For safety, of course, all non-range practice should be done only after you have unloaded your weapon, counted the rounds you've removed to be sure they're all out, and then *rechecked* to be sure the weapon *really is* unloaded.)

SURVIVAL BRIEFING

Some cops look upon briefing (roll call, or assembly, or whatever else you call it at your department) as a time for the sergeant to act tough and the troops to act blase and make jokes. But when you're getting a briefing on recent dangerous crimes or stolen vehicles, you should look upon this as part of your survival preparation: you're being given *advance information* about the largest category of your survival-threatening adversaries.

The criminals who committed those murders, robberies, burglaries, rapes and auto thefts are out there somewhere. *They're* watching out for *you*. They know how you're dressed. They know what kind of car you're driving. That puts them in a position of superiority as to adversary identification. To minimize your built-in inferiority in this area, you need to know all you can about *them*. So when the sergeant is briefing you on everything that's known about your potential killers out there, *pay attention. Make notes.* This isn't just an enforcement briefing—it's a *survival* briefing.

VEHICLE CHECK

We talked about the traffic safety check of your vehicle earlier. In addition to that check, you want to take half a minute or so to do these things at the start of the shift:

- ☐ *If car is in an open parking area, accessible to the public, walk around and visually inspect the car before opening doors or trunk lid.*
- ☐ *Look for strange wires, pieces of tape, string, alligator clips, or other unusual objects dangling from the car or lying on the ground.*
- ☐ *Check trunk, hood, doors and gas cap for signs of tampering.*
- ☐ *Look through the windows for wires dangling from the dash or radio.*
- ☐ *If you see nothing out of the ordinary, open the door and look, then feel, beneath the driver's seat.*
- ☐ *To be sure a previous prisoner hasn't hidden a weapon in the car that your next prisoner may find and use against you, check beneath the front seat by feeling from the rear; check under rear floor mats; pull the rear seat up and check beneath and behind it for guns, knives, razor blades, brass knuckles, etc.*
- ☐ *Check the shotgun and rack. Be sure the switch opens the lock, and that the shotgun is clean and loaded (magazine, not chamber).*
- ☐ *Before leaving the station, make a radio check.*

KNOW YOUR LOCATION

With proper survival consciousness and motivation in the front of your mind, protective equipment in place, an "enemy intelligence" briefing under your belt, serviceable weapons and ammo available in addition to other basic gear (baton, mace, cuffs, etc.), and a safe vehicle surrounding you, you're ready for the street.

From the moment you leave the parking area until you return at the end of the shift, you've got to know from one minute to the next where you are. Every time you turn onto a street, glance up and see what the street name is. Every time you drive into the next block, glance at the curb or building addresses to see what hundred block you're in. Every time you roll past a highway exit or intersection or prominent landmark, make a mental note of it.

If you suddenly come under fire and need to call for help, you can't call "time out" while you run back to the nearest street signs to find out where you are. If you call in the *wrong* location or *no* location, frenzied officers are going to be running around all over the place looking for you, putting themselves in increased danger, and maybe not finding you until it's too late to do anything except scoop you up.

Don't rely on your partner to keep track of your location. When the crisis unfolds, your partner may not be able to tell you anything. Whether you're the driver or the shotgun rider, it's *your* responsibility to know exactly where you are, at all times.

SURVIVING FIELD THREATS/139

"MY LOCATION? I'M . . . I'M . . . ?"

It doesn't do you much good to call for help if you don't know where you are. Always be aware of your location.

FIELD INTERVIEWS

Of the 93, 8 officers were slain while investigating suspicious persons or circumstances. During the preceding ten-year period, a total of 88 officers met their deaths in such circumstances, accounting for 1 of every 12 officer homicides.

When you want to FI a pedestrian on the street, take a few seconds to think over the survival aspects of the stop. Would it be better to talk to him where he is now, or at some location ahead on his apparent route? Do you need a back-up in the area before you tip your hand? Can you get on the air and call in the stop before it becomes clear to the pedestrian that you're interested in him?

In selecting the location (when circumstances permit), favor a spot that has high public visibility (busy street, busy parking lot), over one that would let a suspect act without fear of witnesses (alleys, side streets, remote areas). At night, prefer a well-lit area to a dark one.

Try to pick a spot that limits the suspect's escape routes; open fields or areas blocked off with high fences, walls or solid buildings are better locations than sidewalks adjacent to apartment house complexes, rows of houses, or wooded areas. If you have reason to suspect that the suspect is armed or dangerous, don't stop him near areas with a high risk of bystander involvement (either as hostages or victims of flying bullets). In such cases, avoid parks, playgrounds,

SURVIVING FIELD THREATS/141

A poor location for an FI, giving the subject cover and escape options.

A good FI location.

schools, crowded intersections, or other areas where crowds are concentrated.

Consider the availability of cover for both the suspect and yourself. Ideally, you want to time the stop so that the suspect is caught in the open, and you are surrounded by good cover.

Don't stop the suspect until you've called in the location of the stop. You don't want to be sitting in your car, or have your gun hand filled with a microphone, when you let the suspect know you want to talk to him. First, pick the best available spot—either where he is or where he's going, decide whether to request assistance (depending on your reason for stopping him, his appearance, time of day, and location), and call in the stop, before you call to the suspect or approach him.

I often see the advice given that you always want to approach the pedestrian from his rear—presumably so he won't know you're there. One survival book suggests that if you're approaching from his front, you pass on by, get behind him, and then initiate the stop. I disagree.

Unless you're working plainclothes, you have a hard time being invisible. No matter how disinterested you act as you walk or drive past someone who's got something to hide from the law, he's going to be overcautious to the point of suspicious paranoia at the mere fact of your presence. For all *he* knows, you've made him for a crime that *you* may not even be aware of. If you meet

SURVIVING FIELD THREATS/143

The danger of letting a pedestrian talk to you while you're seated in your car. NEVER let anyone put you at this kind of disadvantage, just because you're too lazy to get out.

him and go on past, you're turning your back on someone who may use it for target practice. *Never offer your back as a target!*

What's more, anytime you approach someone from the rear, his hands are too easily concealed from your view. You can't tell if his hands are reaching into a front coat pocket, or inside the waistband, or inside the coat to a shoulder holster. If he's doing any of these things while you're walking or driving up from his rear, he can spin around and shoot you quicker than you can say: "Oh, no!"

I recommend that whenever possible, you approach an FI subject from a position which *never* shows him your back, and which *never* puts his hands out of your view. I also recommend that you try to engineer the stop so that you keep your patrol car between him and you as you're getting out—even if this means stopping on the wrong side of the street, or pulling into a driveway across his path. If he unexpectedly opens fire when you open your door, you want the engine block between *his* muzzle and *your* vital organs.

Unless you have a handy-talkie on your belt, hang your microphone outside your car door when you get out, so you can get to it and use it without having to bend inside the car window or door. If you're going to be standing on the passenger side (non-traffic side) of your car during the FI, put the mic out that side.

Before you call out to the pedestrian, position yourself behind cover, quickly survey his clothing or carried articles for clues as to whether he's armed (excess clothing, or a coat draped over an arm, should

SURVIVING FIELD THREATS/145

Try to put your engine block between you and the FI subject when you stop, keep his hands in view, and don't turn your back to him.

arouse your suspicions), and depending on the strength of your suspicions about him, either have your gun hand ready to unsnap the retainer strap on your holster, or have the strap unsnapped with your gun hand on the grips and ready to draw, or have your gun drawn and held down out of sight. If you believe the pedestrian is an armed fugitive or recently commited an armed crime, get your weapon trained on him right away.

Since most of your FI's will be investigatory, with little or no reason for suspicion that the subject is armed and dangerous, you should not start the conversation with an authoritarian challenge (such as: "Halt! Police! Don't move!"), which sets a high level of tension and makes a safe, productive FI more difficult. Instead, say something like: "Hey, can I talk to you for a minute?" You can always escalate the firmness with a more demanding tone, or even a verbal command, if the subject doesn't respond to your "friendly invitation."

As he approaches your position, try to remain behind cover until you can see his hands. When he gets close enough to see, if his hands are in his pockets, ask him to bring them into view, slowly. Keep your gun hand poised until you can see that he isn't holding a weapon. Holster your weapon and snap the safety strap before leaving cover to meet him. *Keep you eyes on his hands throughout your contact.*

During the FI, stay far enough away from the subject that he can't reach you with a scissors kick (about 4 feet away). Keep your non-gun side turned to his front,

SURVIVING FIELD THREATS/147

Degree of firearm readiness depends on the situation.

Ready to draw.   **Drawn, down out of sight.**

so that he has no chance of suddenly leaping forward and grabbing for your gun, and also to protect your vulnerable spots (face, throat, solar plexus, stomach, groin, kneecaps, shins and insteps) from a jab or kick. Don't allow the subject to close in on you—even if you have to step back and tell him to hold it. Don't allow him to walk around to face you or to get to your rear. Turn your body as he changes position, so that you continuously present only your non-gun side to him. *Stay alert to your nearest cover possibility.*

While filling out your FI card, glance up at him frequently. Anytime you see his hands moving in your peripheral vision, look up and watch him. Don't let him reach into areas from which he could draw a knife or gun, except under your direction and control. In those cases, don't let him reach *across* his body—direct him to remove items from his right side with his right hand, and from his left side with his left hand. Always be prepared to throw your FI cards in his face, duck behind cover, and draw.

Don't let third persons walk up behind you during the FI. If possible, keep your back to a solid wall or your car to protect your rear, or move yourself and your FI subject. Don't let yourself get placed between 2 people on the street.

When you return to your car or motorcycle to run your subject for record and wants, *keep his hands in view* continuously, even if you have to direct him over to another spot (but don't let him get cover superiority on you). If necessary, walk backwards or sideways to get to your microphone. *Do not offer your back as a*

SURVIVING FIELD THREATS/149

Don't expose your sidearm this way.

A good FI position, with gun side and rear protected, and subject more than kicking distance away.

target. Use your non-gun hand to operate the mic, and keep your gun hand free and ready. Glance frequently at the subject's hands while running the warrant check.

If you're working a 2-man unit, one of you should conduct the FI and watch the subject while the other runs the warrant check. Don't let the subject get between you and your partner, making you unable to fire without endangering each other. I recommend that one of you remain near the car and radio, behind cover, while the other confronts the subject. It is extremely unlikely that he will engage one of you in a shooting if he stands no chance of getting you both.

When you're working alone and stop to FI 2 or more pedestrians, try to keep them in a single line (shoulder to shoulder) and follow all the same precautions as for a single subject. When you have a partner and stop 2 or more FI subjects, I still recommend that one officer confront and FI, while the back-up officer remains behind cover, with the microphone, so the subjects still have no good chance of getting you both with a sudden attack. The best protection for the exposed FI officer is to have the back-up officer in a superior covered supporting position, minimizing the temptation for the suspects to take offensive action.

When the FI is complete, let the subject go on his way. Remain near cover and watch until he's out of the area before dropping your guard and getting back into your car.

SURVIVING FIELD THREATS/151

Don't let anyone walk up from your rear during the FI.

Partner officer should remain behind cover, with radio access, to deter aggressive action by the subject.

CARSTOPS (NON-FELONY)

Sixteen percent of slain officers were killed during traffic stops or pursuits, while enforcing traffic laws. That's the same percentage as lost their lives while investigating robberies in progress, or pursuing robbery suspects. In fact, only one other category of adversary activity—attempting arrests—proved more deadly for the 93 officers than the traffic stop. Even though the traffic stop is one of your most routine activities, you should consider it one of the most *dangerous*.

Many of the same safety considerations of the FI stop apply to the carstop. You want to be able to get on the air and call it in, and get a back-up rolling, if necessary, before turning on your colored lights to pull the car over. If speeds and circumstances permit, run the plate for local and NCIC wants before initiating the stop. If the car is hot or the owner is a fugitive, you want to know that before making the stop, so you can treat it as a felony carstop (discussed later).

If the car is already parked, you have to take it where you find it, unless you can direct the driver, over your outside PA speaker, to move a short distance to a better location. If the car is in traffic when you first decide to stop it, as is usually the case, take speeds into consideration and project where the car will be by the time your radio call is complete, calling in the location where you intend *to make* the stop, not where you happen *to be* at the moment you get on the air. If the car stops short of or beyond the place you've called in, correct your location with the dispatcher so that help

SURVIVING FIELD THREATS/153

If at all possible, do not make a carstop near a schoolyard, park or playground.

A good carstop location, limiting escape access and bystander risk, while affording adequate public visibility.

can find you quickly, if necessary.

Once again, you want to select a location that gives you control, public visibility, sufficient room to avoid blocking traffic, good cover for you, and scant cover for the occupants of the car. If it's nighttime, you want overhead street lighting, if any, to be just *in front of* the car you're stopping, so that the occupants are silhouetted for you. Do not stop with the light just above or behind your car, which would make you a silhouette target for the people you've stopped.

At night, shine the high beams of your headlights on the car ahead once you've both come to a stop, to illuminate the interior. Use your white spotlight back and forth across the car's interior to allow you to see how many people are inside, and what they're doing. Then park your spotlight in the driver's outside rear view mirror, so he can't use it to watch your approach, and so his night vision gets ruined when he looks into the mirror. If you have a second spotlight, park it in his inside rear view mirror, or in the outside mirror on the passenger's side if there is a passenger in the car.

Before you get out of your car (day or night stop), take a few seconds to look at what you've got (but don't stay seated in your car if the driver or passengers of the stopped vehicle are getting out and coming back toward you). Look to see if the occupants of the car are reaching or moving around. They may be looking for license, ID or registration, they may be trying to hide their dope or alcohol, or they may be concealing or withdrawing weapons. Although movement by the driver or passengers will usually be for one of the first 2

SURVIVING FIELD THREATS/155

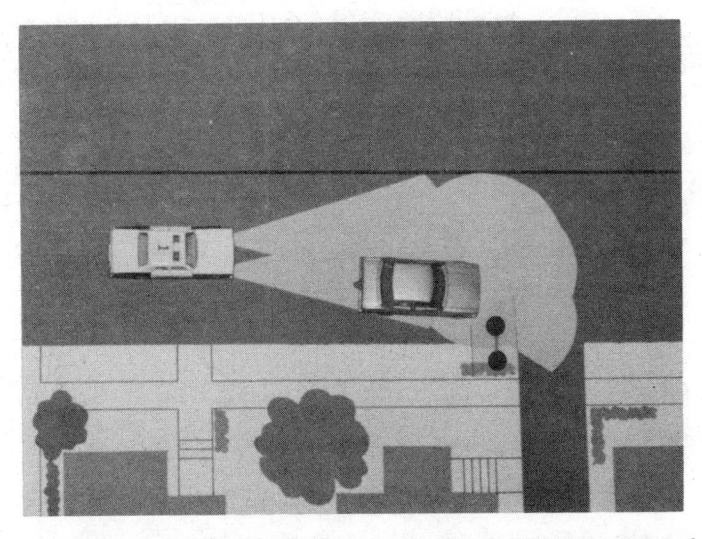

At night, time carstops so that occupants of the stopped car will be silhouetted by street lighting.

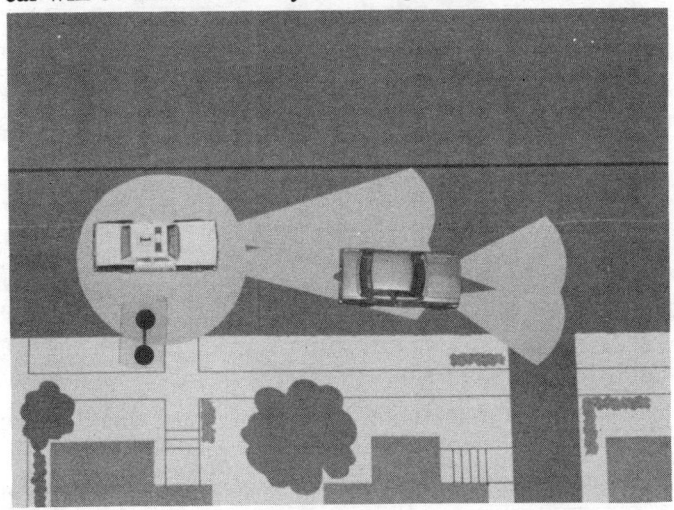

Don't stop your own car beneath a street light.

reasons, you should always act as if observed movement indicates the presence of weapons. Call for a back-up, if warranted.

Besides activity, look for number, age and seating configuration of the occupants. A full carload of young men or juveniles might, depending on other facts, suggest a youth gang battle wagon, which could be loaded with all kinds of weapons. A 4-door car occupied by a driver in front and a single passenger in back suggests a possible robber-driver team. A driver or passenger who ducks down out of sight and does not come back up into view is especially suspicious—and probably dangerous.

Scan the car itself for clues of unusual activity. License plate irregularities (no plate; attached loosely; clean plate on dirty car; dirty plate on clean car) suggest a stolen or getaway vehicle. Signs of trunk-lock tampering, such as a punched lock, may suggest a stolen car, or an ambusher concealed in the trunk. Movement of the trunk lid should be viewed as a clue to imminent ambush.

Fresh body damage may mean the car was recently involved in a hit-run accident, which could make the driver spooky and dangerous. A brand new car without plates or wheel covers, and with a thin coat of dust on the body and a manufacturer's price sticker on the window, may be stolen.

Anything about the car that's unusual could mean you're walking up on more than a traffic violation. Call for a back-up and treat the stop as a felony carstop if you have any doubts at all about your safety.

"SURPRISE!"

Before you leave the cover of your car, scan the trunk area of the stopped car for indications that this kind of deadly surprise is in store for you. If you get careless and let this happen to you on a "routine" traffic stop, you're dead.

If nothing appears to be out of the ordinary, hang your mic out the appropriate window (usually, the non-traffic side). Cut your wheels at a 45 degree angle toward the side of the car you'll be getting out on (usually, the driver's side) to give your feet and legs some cover against low fire or ground ricochets. Carry your ballistic clipboard or metal citation case in your non-gun hand during daylight stops. At night, when your flashlight is carried in the non-gun hand, leave your ticket book on the front fender of your car.

For night stops, approach the stopped car *within* your safety corridor, but *outside* the direct line of your spotlight. If an approach on the passenger side is appropriate, walk around the rear of your car, or slide out your passenger door, to get to the right side. Do not walk in front of your car, through the "leg smashing" danger zone, where you would be silhouetted at night and could be crushed if the driver suddenly accelerates in reverse.

As with the FI, your degree of weapon readiness depends on the degree of danger you perceive. Poise your hand, clutch the grip, unsnap the safety strap, or draw and hold your weapon down behind your leg as you approach, depending on the situation. To avoid accidentally shooting your toes off, keep your finger outside the trigger guard, lying along the frame.

Just before you pass the cover of your front fender, quickly survey the trunk lid and lock for signs of trouble. If in doubt, feint a step forward, then recover

SURVIVING FIELD THREATS/159

When you'll be working from the passenger side of your car, hang the mic out the window so you can get to it without bending inside, and cut front wheels to protect your feet and legs.

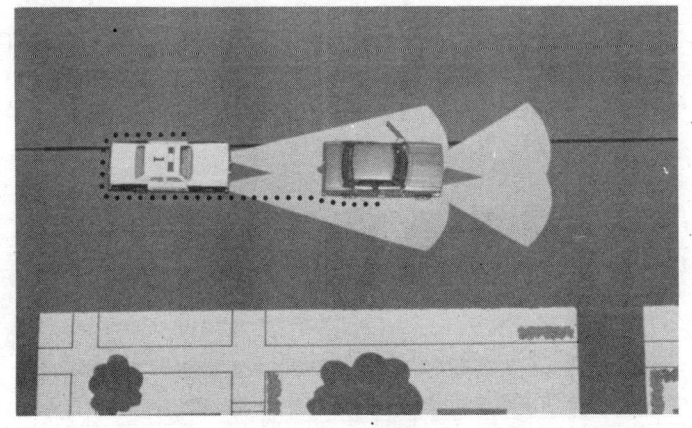

During night stops where a passenger side approach is needed, slide out your right door, or walk behind your car—never in front of it.

behind the fender while glancing from the trunk lid to the interior of the car. If no signs of trouble arise, step quickly alongside the rear quarter of the stopped car, looking through the rear windshield for concealed rear passengers. Stay alert to your nearest cover possibility.

If the back seat is clear or passenger activity seems normal, approach the front door, staying in close to the car, and stopping 3 or 4 inches short of the door opening. Conduct your conversation with the driver from this position, so as to avoid the danger of his suddenly pushing the door open and knocking you into traffic. Standing slightly to his rear also makes it more difficult for him to point a gun at you.

Day or night, look first for the driver's hands, and then for the passengers' hands, while you're telling the driver why you stopped him. If anyone's hands are concealed, tell them you'd feel more comfortable if they'd rest their hands on their knees (driver's hands on steering wheel). Be ready to draw as concealed hands come into view.

At night, you next want to shine your flashlight directly into everyone's eyes, one at a time, to destroy their night vision. You can pretend that you didn't mean to do this ("Oh, I'm sorry . . . Oops, got you, too . . . Sorry").

Day or night, once you see that no one is holding a weapon, you want to quickly scan the interior of the car to be sure no visible weapon is available to the occupants. Your eyes should dart to the seat beside the driver (and passengers, if any), then to the center console, then to the floorboard, then to the dashboard,

SURVIVING FIELD THREATS/161

Standing in this position makes it easy for the driver to knock you down by opening his door, or to aim a gun at you.

This position gives you greater protection.

then to the sun visor, then to any bags or clothing on the seat or floorboard within arm's reach.

If the engine is running, ask the driver to turn off the ignition. If you're going to write the driver or FI the passengers, I recommend you have them get out, one at a time, on your side. *You* should open the door, so the driver's hands don't drop out of sight. Stay back out of arm's reach, with your gun side turned away. Move the driver and passengers to a safe spot off the road. Weather permitting, ask the passengers to sit on the curb with their hands on their knees. Or ask them to stand at the right rear of their car, with their hands on top of the trunk lid. Take the driver or the FI subject back to the area of your microphone to write the citation or FI card, while awaiting reply to your record check.

If you have a partner, he can watch the passengers while you write the ticket or conduct your interview. Avoid crossfire positioning.

Throughout your contact, remember to keep your distance (at least 4 feet), protect your gun side and areas of vulnerability, and do not turn your back to any of the subjects. If control becomes a problem, or a car search is indicated, or an arrest is imminent and you are outnumbered, request assistance and wait for it. Don't try to search a car with the driver outside behind you. If no back-up unit is available, pat the driver down and ask him to sit in the back seat of your car (if you have a cage) until you've finished looking through his car.

When opening the suspect's trunk, stand off to the side, stay low, use your weak hand on the key, and be

SURVIVING FIELD THREATS/163

If you were making this carstop, how long would it take you to spot the "piece" on the dashboard?

prepared to draw (or have gun already drawn, depending on the circumstances). If you have arrested the driver, or with his permission, you can place him between you and the trunk as a shield while you open it.

Some survival instructors suggest that you leave driver and passengers in their car, and you get back in your car to write the citation. I don't like this procedure, for several reasons. It forces you to make your approach to the car *twice,* and most officers are going to think that after walking up in safety the first time, there's no need for a cautious approach the second time. That's dangerous.

Moreover, while you're sitting in your car writing out a ticket, it's hard to keep track of what the occupants are doing. They have plenty of time to conspire together without being overheard, and to pull weapons out of glove compartments, handbags and inside clothing, and from beneath seats.

While they're still inside, they can start the car and speed away. That's far worse than having them split on foot after you've emptied the car.

And finally, if you're sitting in your car writing a citation at night, you have to turn on your dome light or dash light to see, and then you're a sitting duck. For all of these reasons, you're safer—day or night—by getting the occupants *outside* of the car where you have better observation of them, and they have limited access to places where weapons are often stored.

Vans and recreational vehicles limit your ability to tell how many people you're dealing with, and they make great places for armed suspects to hide. For

TWO WAYS TO OPEN A TRUNK

Once you're into this position, let the trunk lid all the way up, quickly. Raising it slowly will give an ambusher a chance to locate you and take aim.

As soon as you turn the key and release the latch, grab ahold of the arrestee's arm with your weak hand, so that he can't duck aside as the trunk lid springs open. (Note: This officer shoots lefthanded.)

non-felony stops, approach them the same way you approach cars, watching for door movement, a door ajar, or reflections from the windows. After you remove the driver, keep him between you and the vehicle, so that he remains in the line of fire. To open van or RV doors, stay low and to the side, or if possible, use the driver-shield method.

For a non-felony carstop where you have a partner or back-up officer, the standard tactic is for the police driver to approach the stopped driver, while the partner officer or back-up officer stands to the right rear of the stopped vehicle (usually, up on the curb), watching the occupants through the rear windshield. I think this, too, is a poor choice of tactics.

In most instances, the observing officer is standing right out in the open, several feet from any cover. It doesn't take much coordination for a determined driver-passenger team to open fire simultaneously out their windows or doors at the approach officer and the back-up. And if this does happen, the back-up officer is in a poor position to return fire, for fear of hitting his partner on the opposite side of the car.

Another recommended approach, with which I also disagree, is to have the partner officer flank out to the right side and remain concealed behind cover, silently watching events at the car, and trying to avoid detection by the occupants.

As I said earlier, the greatest deterrent value of the back-up officer is in letting the suspects see that there's no way of getting both officers in an initial attack. If they think they can get one officer, but not the other,

This traditional position for the partner or back-up officer provides no cover against a sudden gunshot from the passenger window.

This better position puts the second officer behind some cover, gives access to the radio, and provides greater deterrence against an attack on the first officer.

they're unlikely to try anything. I therefore recommend that the partner officer take cover behind the door and windshield pillar of the patrol car, with his feet inside or protected by the 45 degree angle of the front wheels, and his weak hand on the microphone. If trouble starts, he can call help and return fire simultaneously, without putting his fellow officer in the direct line of fire. Seeing the back-up in this position, the suspects are less likely to open up on the approach officer.

To improve the partner's position, a 2-man patrol car can position the stop so that additional and *better* cover is immediately available next to the partner's door, by stopping adjacent to a utility pole, large tree or mail collection box.

When you've completed the citation or FI, watch the driver and passengers as they return to their car, but be making your way to a position of cover. Let them drive away first, and then you can more safely make your log entries, talk on the radio, and get back into traffic.

* * *

It would be unrealistic to expect that you're going to follow *all* of these procedures with *every* FI or carstop you make. Weather conditions and the degree of perceived danger may dictate modifications some of the time. However, bear in mind that it's often the *unperceived* danger that proves fatal to the unsuspecting officer. Therefore, always assume that there's more to an encounter than meets the eye. Always assume that

Whenever possible, and especially in dangerous situations, stop the patrol car adjacent to items of cover, such as this large steel trash dumpster.

there's an unperceived danger, until you have safely satisfied yourself that there is not. *Even then,* protect your rear, protect your gun side, be cover-conscious, and *stay alert!*

It may sound like an impractical task to keep all of these safety considerations constantly in mind. It isn't. It will probably take you only about 2 weeks of conscious effort to program them into unconscious habit. Your subconscious is accustomed to looking out for your safety.

For example, when you drive a car down the street, you don't have to make conscious decisions about pressing down or letting up on the accelerator, or when to brake, or when to change lanes. You can drive all over town, carrying on a conversation with your partner at the *conscious* level, and yet safely maneuvering an automobile in traffic, monitoring the radio, and checking license plates on the cars you see, all on the *unconscious* level.

The first couple of weeks while I was walking through the jungle in Viet Nam, I constantly asked myself: "If they're going to hit us now, which way will they come from? Where's my best cover?" And I constantly made decisions: "If they come from the right, I'll jump behind that tree over there; if they come from the left, I'll use that little dirt mound."

After a couple of weeks of this conscious programming, my subconscious computer took over the job of risk-evaluation and decision-making. I found myself able to do other things at the conscious level (reading the map, talking on the radio, directing my point man and flank security, etc.), while remaining constantly aware of the risks and cover options presented by the changing terrain. Every time we came under fire (which was about 10 times a week), I knew where to go for cover without having to think about it.

If *you* consciously practice good safety habits, they will become automatic for *you*—you'll keep your gun and your back protected, stay out of arm's reach, and know where the closest and best cover is at all times—all without having to think about it.

This doesn't mean that survival can be relegated to your subconscious mind—when you're facing an unusual or unknown situation or any perceived danger, your survival demands your closest *conscious* attention. The value of the automatic, *subconscious* survival habits is in keeping you protected from the *unexpected* dangers that too often kill officers during "routine" activities, when the officer's conscious attention is focused elsewhere.

FELONY CARSTOPS

The felony carstop is one of those "perceived danger" situations requiring conscious survival tactics. When you have reason to believe you're dealing with a dangerous person, you have good reason to take extra precaution.

If help is available, you should never attempt a felony stop by yourself. Even if there's just a lone driver in the suspect vehicle, and even if he's a 16-year-old "kid," call for some troops. And if you can safely follow the vehicle until help arrives, hold off on the stop til then.

Picking the location for the felony stop is much the same as for any other stop—you want to cut off escape access and give yourself superior cover and observation capabilities. One difference is that you should act on the assumption that a gun battle will take place, and avoid populated areas and busy streets where pedestrians and vehicular traffic might interfere with operations, or be in the fields of fire. Assume that bullets will be flying in *all* directions, and try to maneuver the stop to a place where you won't have to worry about innocent bystanders during the operation (undeveloped areas are best).

When you're ready to stop the vehicle, make maximum use of your outside PA speaker to direct the driver exactly where you want him—stopped in the *middle* of the street. If the driver pulls over to the right curb, as he normally would when you turn on your red or blue lights, he and his passengers (if any) will be closer to the cover of utility poles, trees, fire hydrants, mail boxes, driveway curbs, signposts, ditches, walls and structures; they will be closer to the concealment offered by bushes, shrubs, grass and the shadows cast by objects; they will be closer to any avenues of escape that may exist just off the road (buildings, woods, standing crops, etc.); and they will be closer to sidewalk pedestrians whom they can grab for hostages.

Stopped in the middle of the street (or at least near the center line), the occupants will be isolated. They would have to move across open space in order to reach

FELONY CARSTOP

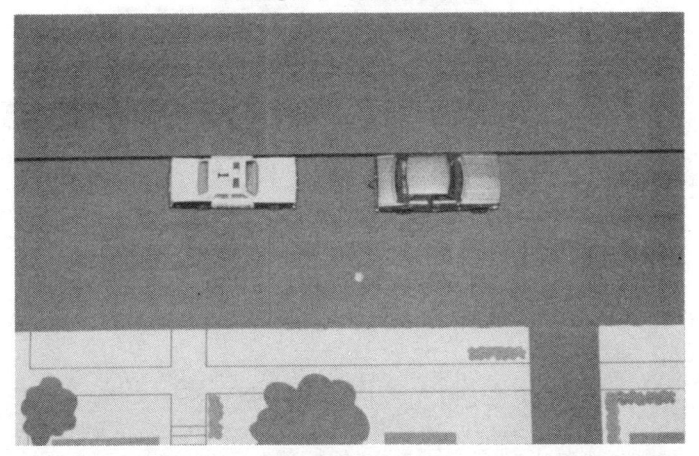

If you can get the driver's attention with your outside speaker, stop him in the middle of the street, away from cover, and you have him isolated.

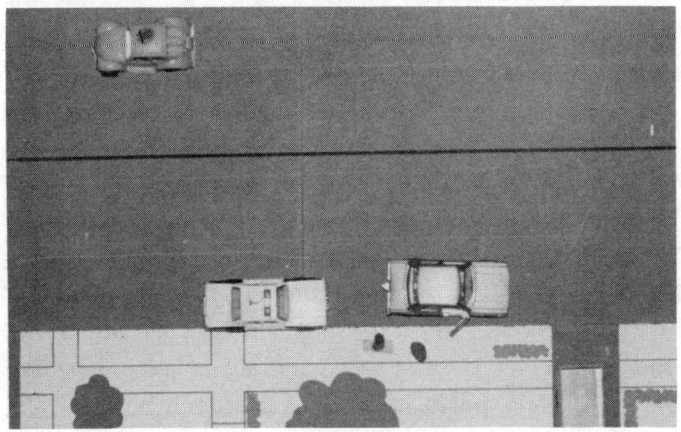

Allowing the driver to pull over to the curb puts him closer to cover, concealment, hostages, and escape routes.

additional cover, concealment, escape routes, or hostages. Therefore, *before* you ever turn on your overheads or spots, honk your horn to get the driver's attention, and immediately announce your orders over the PA: *"Driver in the Camaro—slow down and stop your car—stay in this lane—don't pull over to the side!"*

Start these directions well before he gets to the place where you want him to wind up. You can't very easily back him up, but you can easily direct him forward if he stops too soon: *"Pull forward 3 more car lengths and stop!"*

If you don't have a PA speaker, or if the driver doesn't react to it, you'll have to use your emergency lights (siren if necessary), and make a curbside stop in the safest spot you can find.

Once he stops, you want to position your car about one car length directly behind him. Do not use the offset safety corridor position, because you are not going to approach the car on foot, and you want your headlights and spotlights to be able to illuminate his interior and the area along both sides of his car. (Even if the stop is in daytime, you will still stop in line with the vehicle.) Leave your motor running.

Open your door, but don't get out. If you have no partner, get your shotgun; in 2-man units, shotgun officer takes it. Slide over to the edge of your seat, where your head and upper torso are protected behind the windshield pillar. Stick your shotgun just outside, rack the action to chamber a round, and aim it at the vehicle. Keep your feet inside. Cut your wheels to a 45 degree angle.

SURVIVING FIELD THREATS/175

If you're a one-man unit, put yourself and your shotgun into this position while you give orders to the occupants of the stopped car.

Proper positioning for a two-man unit. These positions provide much better protection than kneeling outside, behind the car doors.

Using your PA speaker (or shouting, if you have no PA), order the driver to roll down his window, so that the remainder of your directions can be clearly overheard by all: *"Driver—roll your window all the way down!"*

Once the window is down, turn off your PA and shout all subsequent orders, unless traffic noise is too high for you to be overheard. Once you switch from PA to voice commands, you no longer have to hold a microphone. Your freed hand can then get to the shotgun, giving you better control of it.

If you're the driver of a 2-man unit, your partner will be in a similar position on his side of the car, door opened, and shotgun leveled at the suspect vehicle. If better cover is available to him near his door, he should get behind it as quickly as possible after the suspect vehicle has come to a stop.

At night, use your headlights and spotlights as described for non-felony stops. Turn off your rotaters and forward flashers to eliminate distracting reflections from the rear windshield of the suspect vehicle. Leave rear flashers on to warn civilian traffic.

You next want to get the occupants' hands into view, and if someone is lying down or bent forward out of view, you want them to think you know they're there: *"Everyone inside the car—sit up in the seat! Driver—stick both of your hands outside your window. Front seat passengers—put both hands flat against the middle of the windshield, fingers spread! Stay away from the sunvisors! Back seat passengers—scoot forward and put your hands on the top of the front seat backs with your palms facing up! Everyone keep your hands there and don't move them until I tell you to."*

"*Driver—keep your left hand in place, and turn off the ignition with your right hand! Throw the keys way out your window! Keep both hands outside the window!*"

If you're waiting for a back-up unit to arrive, hold everyone in this position until the back-up gets into place.

Because of the danger of being caught in conflicting fire, the back-up unit should not approach from your front. Parking position for the back-up unit, depending on street width, traffic, and relative positions of the cars, should be at something between a 30 degree and 60 degree angle from your car, on the side where occupant removal is to occur (usually, driver's side). Terrain permitting, the second back-up unit to arrive should park at a 30 degree angle on the *opposite* side, so both sides of the suspect vehicle are well-covered.

All back-up units should turn off rotaters and forward flashers, and direct highbeams and spots onto the side windows of the suspect vehicle. Back-up officers should assume covered positions behind their windshield posts, or deploy behind nearby cover (within the 90 degree angle), and train shotgun and sidearms on the occupants of the suspect vehicle.

You next begin to remove the occupants through the doors on the side of their vehicle which your force has covered (usually, the driver's side). You want to remove, search and secure only one occupant at a time, beginning with the driver.

178/THE OFFICER SURVIVAL MANUAL

FELONY CARSTOP

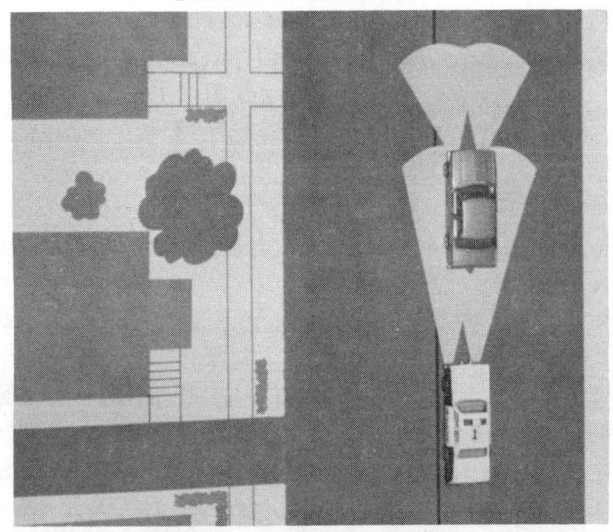

Position of first squad car.

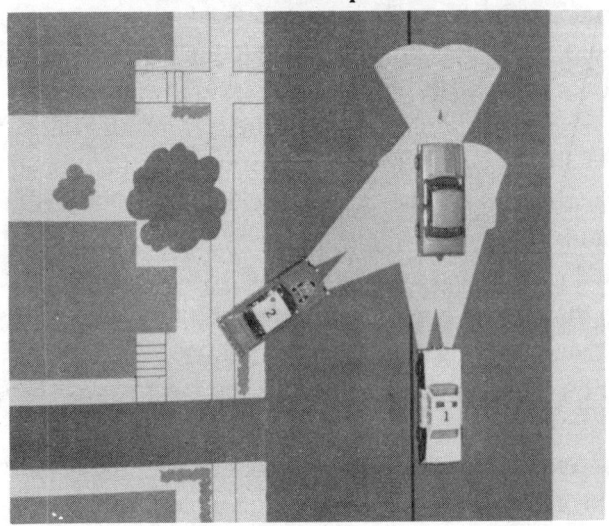

Position of second car.

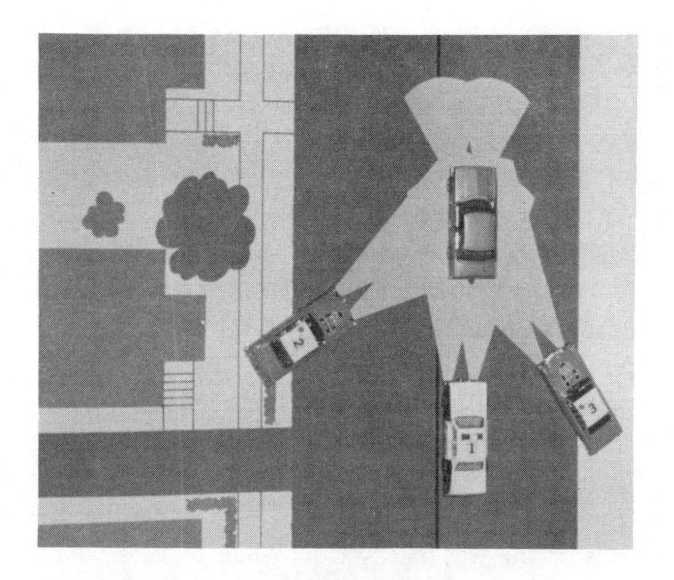

Position of third car.

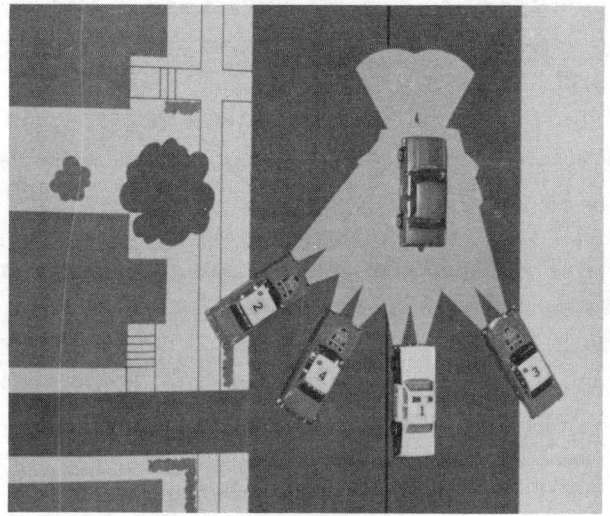

Position of fourth car.

"Everyone inside the car—I'm going to give you some individual directions! Pay attention to what I say! Don't move unless I tell you to."

"Driver—using your outside door handle, open your door! Now, slowly get out with your hands straight up in the air, as high as you can reach! Now, I want you to slowly turn around in a full circle! Keep your hands up in the sky!" As the driver turns, look for weapons or bulges in his clothing.

"Now, driver, slowly walk over here to where I am! Keep your hands all the way up! Whatever you do, don't drop your hands!"

When the driver gets to within 6 feet of your opened car door, stop him and have him turn around and walk the rest of the way backward. When he gets even with your door, order him into the kneeling or prone search position (depending on the weather and ground conditions), as described in a later section, **"Arrest and Control."**

After the driver has been properly searched, cuffed and secured, follow the same procedure with front passengers, getting them out through the driver's door, one at a time. Empty the rear seat passengers from left to right, getting them out through the driver's door (2-door cars), or the left rear door (4-doors).

Once you have all the visible occupants secured, assume there is someone hiding in the floorboard. Tell him you know he's there, and order him out. If there's no response, advance to the rear corner of the car on the opposite side from the opened doors. Have your weapon drawn and following your line of sight.

Peek in through the rear windshield. If you see someone, retreat to the nearest cover and treat the concealed person as a barricaded suspect (although he could be a dead or unconscious victim). Use gas to get him out into the open while officers cover both sides of the car.

If you see no one through the rear windshield, quietly move along the side of the car (opposite from the side with doors open), keeping low. Peek into the right rear window quickly, to see if anyone is in the back seat or floorboard. Be ready to fire through the closed door.

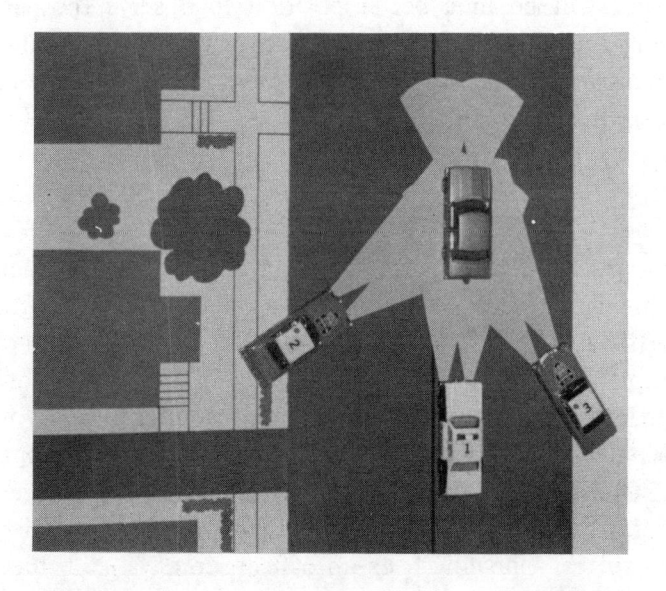

Follow this route to clear the car after all visible occupants have been removed. (A concealed gunman in the floorboard would probably expect you to approach from the *driver's* side.)

Normally, throughout the felony carstop operation, the driver of the initial police car is in charge throughout the operation, regardless of rank, unless an arriving supervisor specifically assumes command. To avoid confusion, only that first officer should give commands. Other officers should concentrate on covering passengers in the suspect vehicle, and assisting in securing detained or arrested suspects.

The van or RV stop may be an exception to this general rule, requiring the officers directly behind the van to provide cover, while officers in the second unit (who will be in a better position to observe the passenger compartment) control movement of the visible occupants. The suspect-shield method can be used to clear the back of the van or RV.

* * *

You may have read other survival books or attended seminars where some of these tactics were taught differently. I'll take just a couple of pages here to explain why my recommendations in some areas are different from the "standard" approach, and then you can decide for yourself which tactics *you* want to adopt.

One traditional theory for placement of the back-up units has been to park them directly *behind* the first patrol car, allowing back-up officers to advance behind the open doors of the lead car. There are several reasons why I think this theory is tactically unsound. For one thing, it limits the amount of nighttime illumination to just what you get from the lead patrol car. Back-up units parked in a column can't throw extra light onto the suspect vehicle; back-up units parked in the 30 degree to 90 degree angle pattern, on the other hand, can flood the suspect vehicle with headlights and spotlights.

Having officers advance to positions behind the open car doors of the lead unit results in having officers bunched up behind inadequate cover. Feet, legs and heads are exposed when the car door is used for cover, and half of the weapons you're likely to encounter can easily shoot right through the car door and into the vital organs of the bunched-up officers. You want as much of the engine and body of your car as possible protecting you. There's no way to give this protection to the back-up officers if they park directly behind the lead patrol car.

Worst of all, you forfeit shooting superiority when all officers approach a fixed target from the same direction; and you give the vehicle occupants the benefit of the back half of their car for protection. If 4 or 6 or 8 officers are all ganged up directly to the rear of the suspect vehicle, a single gunman can effectively tie them all down—he only has to fire in one direction. And the ganged-up officers have to shoot through the back of the suspect's car and hope that a bullet gets through and happens to find its mark. These are poor shooting tactics.

On the other hand, with units parked at angles to the suspect vehicle, every officer has the advantage of better cover in the car doorjamb position; officers are not bunched up, so that a single gunman can't pin them

184/THE OFFICER SURVIVAL MANUAL

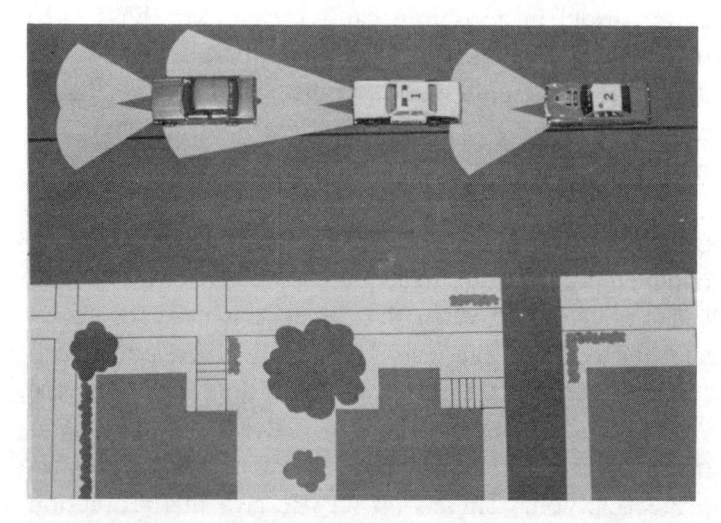

Tactically unsound positioning of the back-up car.

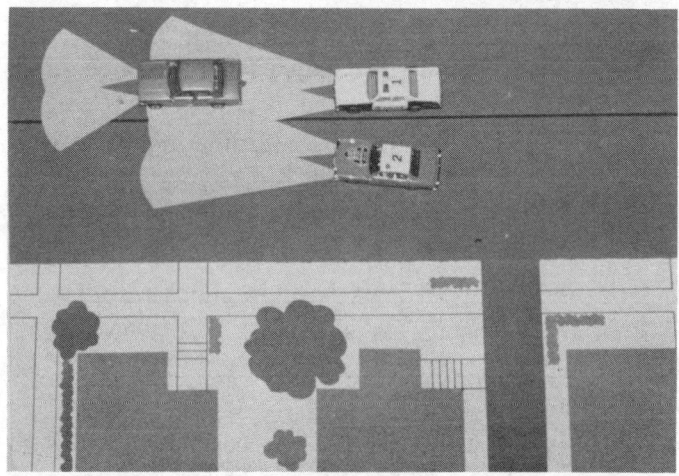

A somewhat better approach, but compare this photo with the one at the bottom of page 178 to see the differences in illumination and firepower patterns.

all down with a single line of fire; and best of all, the officers in the angled car can pour firepower directly into the side of the car, where the occupants will get the least protection from doors and windows. If the operation involves, for example, three 2-man units, you can be delivering firepower into the suspect vehicle from 6 different points—and thus 6 different directions—simultaneously. This makes it hard for the occupants of the car—especially a lone gunman—to return effective fire, or to escape injury or death.

Another standard classroom theory calls for the back-up unit to pull abreast of the primary patrol car. This puts the back-up officers into a better position to assist in arrest and control after the removal of cooperative occupants. This deployment offers better cover, illumination and firepower capabilities than the column, but it's still inferior in maximizing use of lights and firepower into the most vulnerable surfaces of the suspect vehicle, and it still allows a lone gunman too much ease in pinning down a group of officers with unidirectional fire.

For all of these reasons, the angled approach for back-up units is far superior from the standpoint of survival tactics. While I've never had occasion to use this deployment of police cars in a street shootout, I've successfully used the same type of deployment of foot soldiers, tanks and armored personnel carriers in close-quarters fights with isolated enemy concentrations. It works.

Another classroom tactic with which I disagree is the insistence on having the driver of the suspect vehicle do everything with his left hand. Because 87% of people are right-handed, the idea has been to tell the driver to use his left hand to roll down the window and to turn off the ignition and throw out the keys, because, presumably, this would keep his gun hand away from any weapon.

This is another one of those "seminar survival" tactics that sounds logical when you read it in a book, but isn't. Sit in your car behind the wheel, put both hands on the steering wheel, and act as if you've just been ordered to roll down your window. You'll find that you *naturally* use your left hand for this, because it's awkward—if not impossible—to reach across the steering wheel and beneath your left arm to the window crank with your right hand.

If you don't specify use of the left hand when you order the driver to roll down the window, and if you then see him attempting to use the unnatural method of reaching with his right hand, you *know* something's wrong. By trying to do things the hard way, he's just told you there's something (weapon, dope, etc.) he wants to get to with his right hand. You can order him *immediately* to get his right hand back up and use his left hand, and now you have some information you wouldn't have known if you had told him to use his left hand to start with.

By the same token, when you simply order the window rolled down and leave it to him to take the natural choice of using his left hand, his doing so will give you *some* evidence that he isn't inclined to try to take advantage of this open opportunity to do something with his right hand. Again, this is information you preclude if you make the choice of hands for him.

Once you have the driver with his hands hanging out the window (or even if you use the less satisfactory technique of having him place his hands on the windshield or on top of his head), ordering him to reach with his *left* hand to the *right* side of the steering column, beneath the wheel, to turn the key and remove it by pulling it out to the *right,* is forcing an awkward and dangerous maneuver. Some uncoordinated drivers won't be physically able to comply with these directions. Others will have problems. Some will drop the keys on the floor during their efforts to obey, and some of these, not realizing the danger, will bend down to pick them up.

To the nervous cop who's watching the efforts of a frightened, confused driver to perform this series of unnatural, cross-handed feats, the driver may appear to be stalling, or moving around too much, or bending forward to grab a gun. This is no time for unnecessary mistakes. But that's exactly the danger you create by demanding a lot of backward, left-handed contortionism from the driver. What's more, just about anybody who can pick up a gun and fire it with his right hand can pick it up and fire it with his left hand. In my opinion, the risks you create with this chancy approach are not justified by its hypothetical value.

And finally, the current trend in suspect control seems to be having the suspect interlace his fingers behind his head after he's out of the car and walking back to you. My view on suspect control is that you keep the suspect's hands as far away as possible from places on his body where he could have a weapon concealed. Interlacing the fingers behind the head doesn't accomplish this objective, and it often puts the suspect's hands out of your view.

If the suspect is a longhair (which includes most women and many men), he can easily have a derringer in a headband or neck holster. Any suspect could have a knife or gun inside a cap. Any suspect could have a throwing knife in a scabbard just below the collar, inside the shirt or jacket.

When you have the suspect keep his hands straight up in the air, on the other hand, there's no place for him to pull a concealed weapon from, unless one happens to be floating along in the sky. And you can see his hands at all times, no matter which way he's facing or walking. This way, you can keep him turned toward you as he walks from his car back to you, until you're ready for him to turn around. This increases your control, and makes it hard for him to communicate with the passengers in his car.

Incidentally, if you thought that having a suspect interlock his fingers would put him into some form of isometric self-handcuff, try it on yourself. See how quickly you can pull your hands apart. No problem, right? Now, have your partner or your spouse interlock their fingers behind their head and try to keep you from

Ordering a suspect to interlace his fingers behind his head may look okay from the front . . .

. . . but it permits a suspect ready access to concealed weapons, which makes it a dangerous technique.

pulling their hands apart. See what you've got? Interlocking fingers won't make it tough for the suspect to move his hands—it just makes it tough for you to get them apart for cuffing, if he doesn't want to cooperate. All you do when you order a suspect to interlock his fingers behind his head is make it easier for him to reach a concealed weapon, or to resist arrest.

ROBBERY CALLS

Sixteen percent of officer homicides were directly attributable to armed robbery situations, and many of the officer fatalities in other statistical categories may also have involved armed robbers. When you're dispatched to a robbery in progress (including silent alarms which may have been false the last 50 times you answered them), assume that you're going to run into at least 2 armed, dangerous opponents.

Don't rush right up to the front door in your marked car as soon as you get the call. Picture the scene, if you're familiar with it, or look it up on the map. Determine the best way to approach so as to keep out of sight from the establishment. Consider the most likely routes the robbers will take in fleeing the scene, and call for units to cover them. Be sure the dispatcher has assigned you a back-up. Communicate with your back-up officers by radio enroute to the scene, and coordinate your approach.

Depending on time of day, weather and traffic conditions, cut your siren and rotating lights from 3 blocks to a mile or more away from the scene so you don't alert the robbers. In remote areas at night, cut all your forward lights as far back from the scene as you can safely do so.

The closer you get to the scene, the more likely you are to encounter the fleeing robbers. Silent alarm notification to your agency, unless it's direct-wired, usually involves a reporting delay of 1 to 4 minutes; telephone notification usually carries a 2 to 5 minute delay. In many cases, victims won't trip alarms until the robbers are headed out the door. So even though you treat all cases as robberies "now in progress," the odds often are that the robbers will drive right past you as you're making your approach.

Watch for cars coming from the area at excessive speeds. Note the make and color, and if possible, jot down a plate number. Any vehicles coming from the area without plates, with obscured plates, or without running lights at night are especially suspicious. Report their location, description, and direction of travel, and ask the dispatcher to have an unassigned unit intercept.

As you get within a block of the establishment, be especially watchful for someone running—particularly if he's overdressed or carrying something. Also watch for a standing car, occupied or empty, with motor running, or doors open, or backed into parking, or parked irregularly. Report any of these things and have officers assigned to investigate.

Stop your car just short of the establishment, where it can't be seen from inside. You don't want the robbers to know you're on the scene, or they may be tempted to barricade themselves or take hostages.

192/THE OFFICER SURVIVAL MANUAL

Don't drive right up in front of the establishment being robbed.

Park around a corner, or down the street, where your car will be out of sight.

Dismount from your car, taking your shotgun and racking in a round. Move along covered, concealed routes toward the establishment. At the end of cover, stop and peek around to survey the scene. Your partner or back-up officer should cover the rear or side exit, while you cover the main exit. Position yourselves so that neither is in the other's line of fire. Turn customers away, and order them to leave the area immediately.

Then wait. If there's no screaming or shooting going on inside the establishment, don't go charging in like the Marines. You want to wait behind good, solid cover until the robbers come out and pass the point of no return. Don't announce your presence the minute you see the suspects come through the door; if they *are* the robbers, they can simply dash back inside and pick out some hostages. They should be far enough from the exit to prevent this risk before they find out you're there to greet them.

Ideally, you should get them well out into the open, so that if they turn to run back inside, you'll have time to shoot them without endangering people inside. If you have reasonable cause to believe such a person is an armed robber who will endanger lives inside, this is a legitimate shooting in most jurisdictions. Check local statutes and department policy to make sure you have a clear understanding of applicable rules.

Don't let your suspects get too close to their car or other cover before you announce. Have your shotgun (or sidearm, if no shotgun) aimed directly at the suspects, and be ready to fire if the suspects start to take aim at you.

194/THE OFFICER SURVIVAL MANUAL

Don't alert the robber to your presence the moment he comes through the door.

Let him get into the open, away from the business and without any cover, before you order him to halt.

When you're ready to announce, don't waste time or words with some long sentence ("This is the police . . . we've got you covered . . . stop in your tracks . . . "). Just use essential words, and let the tone of your voice carry the message that you mean business: *"Police! Freeze!"* or *"Sheriff! Hold it!"* (Don't draw things out by adding words like "officer" or "deputy." You don't have time to rattle off personnel classifications.)

Once you've made the announcement of your presence, handle the call according to procedures under the topic of **"Arrest and Control,"** if the suspects cooperate, or as per **"Use of Force,"** if they do not.

Be alert to the possibility that robbers may have lookouts and back-up men stationed nearby, either in places of concealment, or nonchalantly acting like disinterested bystanders. Scan the area frequently and quickly. Do not let anyone approach yourself or the suspects. Bring in your back-up officers for area security as soon as the suspects have been neutralized.

Treat the suspects as armed robbers until and unless you verify from the victim that they are not (you may have misjudged customers in either a real or false alarm situation). Make certain of victim identification before assuming that someone who represents himself as the victim really is.

In those cases where you watch and wait for robbers who never come out, use your walkie-talkie or signal a fellow officer to go back to a car and call the dispatcher, who should have verified or unfounded the robbery by this time via telephone. If the robbery still appears to be genuine, call for more assistance and treat the situation as a barricaded suspect case. Do not advance toward the establishment without cover and concealment unless you get independent confirmation that the alarm is unfounded, or the circumstances compel you to draw that conclusion (many customers leaving in normal fashion, and business apparently operating as usual, for example). In the latter event, remain cautious as you advance, enter and confirm that the alarm was accidental.

Don't forget to notify fellow officers who intercepted suspicious vehicles or pedestrians leaving the area, as soon as you have determined that there's no cause to hold them.

BURGLARY CALLS AND BUILDING SEARCHES

Over the years, approximately 7% of officer homicides have resulted from burglary calls and pursuits of burglary suspects. Statistically, burglars are not likely to be armed, nor too dangerous. Those statistics provide no comfort, however, to the surviving families of 76 officers who have been killed by burglars in the last 10 years. Just because the typical burlgar *prefers* stealth to hostility doesn't mean he won't turn hostile when he fears that you're about to arrest him.

Until you have them neutralized, treat all burglars as if they are the kinds of opponents which 76 late officers misjudged.

The approach to the scene of a reported burglary alarm should follow all of the same guidelines as the robbery approach, including alertness for fleeing suspects, and light and noise discipline on close approach.

Two cars covering a building on a burglary alarm call. For best shooting position around corners, the perimeter search on foot should be conducted in a counterclockwise pattern.

Watch especially for vans, trucks and juveniles.

If the form and layout of the structure permits, 2 units should cover the exterior by positioning themselves at diagonally opposite corners, allowing each unit to watch 2 sides of the building. Before leaving their cars, officers should survey the building for suspects on the roof or ledges. On cue, officers armed with loaded (chambered) shotguns should approach their respective corners on foot, making use of available cover and concealment.

Officers should next conduct a perimeter search, moving along the walls and using cover, and avoiding silhouettes against glass doors and windows. Both officers should work their way around to the opposite diagonal corners, moving only in a <u>counter-clockwise</u> direction. This prevents the officers from bumping into each other, and allows them the most advantageous shooting position around corners (except for southpaws). Stop and peek quickly around corners before walking around them.

During the perimeter search, you should look ahead, to your flanks, up and back. Look not only for suspects, but also for a ladder leaned against a wall; a fire escape ladder pulled down; a rope dangling from a ledge, window or rooftop; boxes stacked beneath a window or low overhang; broken windows; damaged, open or unlocked doors; or merchandise stacked outside the building.

If your perimeter search reveals no point of entry or means of rooftop or basement access, maintain security from diagonal corners until the owner arrives to let you in for an interior search, if the circumstances warrant.

If you do find a point of forcible entry during your perimeter search, contact your fellow officer by walkie-talkie, flashlight signals, or voice contact. Don't go in alone at this point. Wait for assistance. Call for more troops if necessary.

Be sure to check out possible hiding places outside the building, in the vicinity of the POE. Using quick peeks around barrels, packing crates, or whatever else is there, make sure the burglar hasn't gotten behind you.

In daylight or in lighted buildings, more than one officer should go in to conduct the building search. Don't enter through damaged doors in such a way as to disturb physical evidence or obliterate possible fingerprints. If necessary, wait for the owner to let you in elsewhere. Don't climb through windows—that's just asking to get shot in the head.

If the door is already standing open, take peeks around the edge (first high, then low) to look for suspects and cover. If the door opens inward, try to use your baton or a long stick to shove the door all the way back against the inside doorstop, to be sure the burglar isn't waiting behind the door. If the door opens outward, have your partner on the back side of the door quickly pull it all the way open.

If the door is closed, but unlocked, when you come upon it, you should be able to tell when you test it which way it opens. If the door is flush with the outside of the doorjamb, it will open outward; if it is recessed in the doorjamb, it will open inward.

Your method of entry will depend on where your closest inside cover is. If it's on your side of the door, stay close to the doorjamb and keep low as you go in. There is a "standard" entry technique called the "wrap-around," in which you start with your back flat against the outside wall, and wrap your body around the door frame, putting you just inside, against the wall. You can try this appraoch in practice and see what you think of it. I've found it to have 2 major flaws: you can't do it unless you're a cousin of Elastic Man, and if you *can* do it, you find yourself *facing* the inside wall at the finish, with your back exposed to anyone inside.

If the nearest cover is directly in line with the doorway opening, keep low and dart straight in quickly. If the cover is across the opening, keep low and dart in quickly on a diagonal line. You want to avoid silhouetting yourself in the opening, and as you move through, you want to present the smallest, fastest-moving target you can.

Once inside behind cover, look behind you, to both sides, ahead, and above. If the coast is clear, signal your partner in.

To search the building, you and your partner should leapfrog, each covering the other's advance to the next place of cover. Clear one floor at a time, searching each room as you come to it, and bypassing no place where the burglar might be concealed. Continue quietly in this fashion, frequently looking in all directions to avoid being surprised, until you have cleared the entire building.

If you locate anyone during your search, assume that he's armed and dangerous. Handle him as per **"Arrest and Control."** Act on the assumption that other suspects may also be in the building, and *don't relax* after securing the first and any subsequent suspects until you've cleared the entire building.

For nighttime searches of a <u>*completely dark*</u> building, I recommend that only one officer enter, while back-up officers remain at the outside door. It's impossible to use a leapfrog advance in the dark very safely or very quietly. Any form of communication between officers in the dark—whether by sound or by light signals—can betray your location to an unseen suspect. You're likely to bump into your partner in the dark, or get split up and put yourself into an uncertain shooting position. You'll be inclined to use flashlights to avoid stumbling into each other, and the minute you flick one on, you become a walking target, and you spoil your night vision.

All of these disadvantages of a 2-man entry are eliminated if you go in alone. You don't have to talk or whisper or snap your fingers or tap out coded signals or blink your flashlight. You don't have to worry about tripping over your partner's feet. You don't have to wonder whether the noise you hear is coming from your partner or the burglar. You don't have to be afraid to return fire. And you don't have to walk around with a flashlight lit up like a sign saying: "Here I am, Burglar. Go ahead and take a free shot."

When you're by yourself, you move by night vision, smell, hearing, and touch—not with a lot of noise and light and confusion. Without a flashlight in your hand, you can use one hand to feel your way along, as you keep crouched low, and you have your hand free to help manage and use your shotgun. It's hard to hold a flashlight in one hand, a shotgun—or even a pistol—in the other, and move through a dark, unfamiliar building without breaking your legs.

I know that some survival instructors teach you to hold your flashlight in your extended weak arm, away from your body, and high and slightly forward. If there's sometime when you absolutely have to use a flashlight in a building search, that's the best way to use it, alright. But most officers use it unnecessarily, and in their attempts to make things easy for themselves, they also make target identification easy for a concealed, armed burglar.

It doesn't take a really bright felon to know that a flashlight doesn't float in mid-air. If you think the average burglar can't figure out that the flashlight he sees coming in his direction has got to be at the end of somebody's non-gun arm, you're underestimating your enemy. And if you think you're going to escape the scatter pattern when that burglar pumps off a round or two of double-ought buck at the general vicinity of your flashlight, you're overestimating the length of your arm.

Despite the fact that going in alone on a dark building search deprives you of the benefits of having the assistance of a back-up officer, I think you're much safer posting him just outside the door. One of the nicest things about having the back-up officer along with you inside is the boost it gives your morale, just knowing

WHAT THE BURGLAR SEES

... if you use a flashlight   **... if you don't**

he's there. That's *nice*, but it's not *good*. If you start taking false comfort in the knowledge that your partner is right behind you, you're more likely to start taking risks you wouldn't have taken by yourself. That's not good.

On the other hand, if you go in alone and scared, you're going to be more cautious. You're not going to be making all that noise, or using all that light, or worrying about your partner's location, or depending on his presence somehow to save you once you've clumsily told the bad guy where you are.

If your search will extend to upper floors, by all means go back and get your back-up man after you've cleared each floor and take him with you to the entrance to the next level. This keeps him available when help is needed, and if the burglar should manage to slip past you, your back-up officer will be there at the door to greet him.

(It's tactically better to clear multi-level structures from the top down, if upper-level entry is available to you. However, that's hardly ever the case. Whether you're moving up or down, use the stairs—not noisy, confining elevators.)

If things hit the fan, noise discipline will be broken, your back-up officers will be charging in to find you, and you can shout to them to direct them in. For the amount of time it takes them to reach you, you will have sacrificed firepower as a result of going in alone. Weighing this disadvantage against the risk reduction you achieve by going in alone, my personal evaluation is that you improve survival odds with the solo search *(dark buildings only).*

Before you decide on the technique that *you* intend to use, try out both my "solo" recommendation and the more common "team" approach for darkened building searches (do it under controlled conditions—don't experiment at a real burglary scene). See which method you think is less likely to get you killed, and adopt that one.

Fortunately, most buildings will have sufficient lighting at night to permit any number of officers to enter and search, using the leapfrog method to cover and advance. And, of course, better than trying to conduct a search alone in a darkened building is to <u>turn on the lights and use additional troops</u>.

DISTURBANCE CALLS

All disturbance calls, including family fights, man with a gun, shots fired, and bar fights, continue to be one of the most dangerous types of assignments for officers, accounting for nearly 16% of officer homicides during the last 10 years. Of the 93 officers slain in the year analyzed, 10 were responding to disturbance calls.

Family Disturbance Dispatch. It's usually a member of the feuding family who reports the disturbance and asks for police help. Most often, this reporting person is a woman, who's having trouble either with her husband, her boyfriend, or her teenager. The person who calls to ask for police help is a source of advance intelligence for you about what kind of threats may be involved on the call. Therefore, your survival on family disturbance calls begins with the desk officer, dispatcher or telephone operator who takes the call.

The person who takes calls should be trained to ask the following questions, at a minimum, and to pass on pertinent information to the assigned officers:

- *What is your name?*
- *What is the address of the disturbance?*
- *Are you there now? (If not, where are you calling from?)*
- *What is your phone number?*
- *Who's causing the disturbance? (Get full name and relationship—whether husband, boyfriend, son, neighbor, etc.)*
- *What is he doing?*
- *Does he have a gun or other weapon?*

- *Is there a gun kept in the house? Where?*
- *Besides yourself and (husband, etc.), how many other people are there?*
- *Does your (husband, etc.) know you're calling the police?*
- *Is your (husband, etc.) wanted by the police for anything? What's his date of birth?*
- *Has he been drinking or using drugs?*

This may seem like a long list of questions to be asking a hysterical caller; however, if the desk officer firmly cuts in on the caller's hectic, disorganized report and commands the caller to answer these short questions, it will take less time to get more information than by simply letting an upset caller ramble on for several minutes without telling you the things you need to know. A firm desk officer can get all of this information in less than 2 minutes.

In assigning the call, the dispatcher should let you know as much as possible about what you're up against:

Baker-Fourteen, handle a family 415 at 1204 South Walnut. Wife reporting husband is D+D, smashing furniture. He's not reported armed; however, she advises a loaded hunting rifle is kept in the hall closet. Two children in the home. Husband is aware of the call. We're running 28 and 29 and will advise. Baker-Seven will follow from 22d and Main.

From the standpoint of your survival, this is a much better dispatch than a careless or lazy one, like:

SURVIVING FIELD THREATS/207

"*Baker-Fourteen, handle a family 415 at 1204 South Walnut. Baker-Seven will follow.*"

While you're enroute, the desk officer can run the name and DOB of the drunk husband for records and warrants, and let you know if you're going to be walking up and knocking on the door of somebody who's wanted for armed robbery in the next county. That's much better than your finding it out the hard way.

Baker-Fourteen and Baker-Seven, your 415 subject is Donald Leroy Adams. Flynn County holds an outstanding felony warrant for 211, 207 and 245a. Also, Adams has 2 priors for resisting. Baker-Three will also follow from the station, ETA in about 10.

A good dispatcher can help you survive.

Often, of course, the reporting party isn't going to be able to supply all the information you'd like to have. And record checks won't always be possible before you get to the scene and leave your car. Some record checks, on the other hand, may disclose no cause for alarm, but you may still be facing an explosive and unpredictable situation at the scene. So always treat every disturbance call as a potential threat to your survival, and proceed accordingly.

If someone at the scene of the family disturbance is going to do you harm, there are 4 different times when they can do it: as you *approach* the scene, as you *enter* the residence, while you're *inside,* or while you're *leaving.* Take precautions to minimize your exposure to risks at all of these times.

Family Disturbance Approach. Always assume that the offender at the family disturbance scene knows that you've been called and are on your way to interfere in his "private business." Expect an unfriendly reception.

If the location is a condominium complex or an apartment house, approach and park where you will be least visible to residents until you can get inside the building. If there are two ways to approach, use the less-obvious route. Find out from the manager exactly where the apartment is located.

If the apartment is on an upper floor, use the stairway furthest from the target apartment to get onto the floor. If you have to use an elevator, take it to the next higher floor and come back down by a stairway away from the apartment. If there are 2 or more elevators available, send an empty one to the floor the apartment is on, while you ride another elevator to the floor just above.

If the residence is a smaller building (boarding house, duplex, single-family house, or mobile home), picture the location in your mind while you're enroute, or look it up on your map if you're not familiar with it, before you drive into the area. Determine from the street address which side of the street it's on, and estimate how far it is from the nearest intersection. You want to avoid driving past the house in order to find it, if at all possible. You want to approach from whichever direction will put the house on *your right,* so that your car body will be between you and the house, and you won't have to cross an open street during your foot approach.

If the house is within walking distance of an intersection, park on the side street and cut across neighboring yards to get there, using available concealment and keeping aware of your nearest cover. If this approach is impractical, park your car several houses away (same side of the street), and approach the subject house from an oblique angle, again using concealment and noting cover options. (At night, avoid parking beneath a street lamp.) When a potential adversary inside that house is expecting you, the *last* thing you want to do is pull right up in front and park, or even worse, pull into his driveway. Ideally, you don't want anyone to know you're there until you knock on the door.

210/THE OFFICER SURVIVAL MANUAL

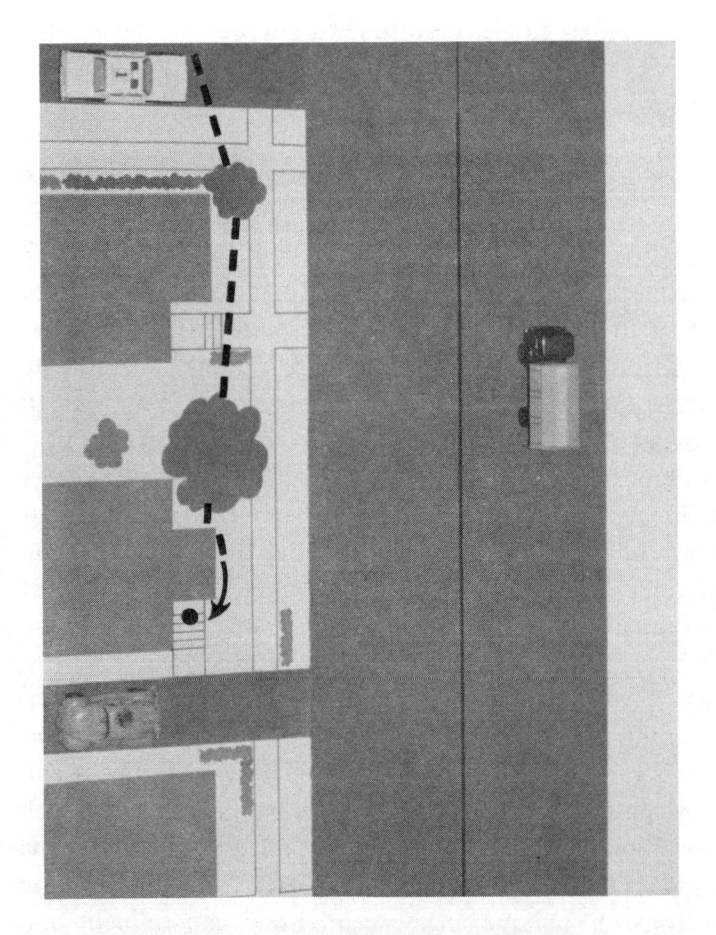

Approaching the family disturbance scene on foot, using cover and concealment to reach the house undetected, is less likely to make a target of you than lazily pulling up in front of the house.

As you walk up, watch not only the windows and doors to the house, but also anyplace where an enraged gunman could be waiting for you: behind plants; inside, behind or beneath parked vehicles; inside or behind garages or outbuildings; or concealed in nighttime shadows. Expect the unexpected.

During your close-in approach, try to avoid moving past windows, or move past them quickly if they're unavoidable.

Entry. Most residential entry doors open inward, from your left as you face them. On all but the most expensive or very old homes, entry doors are usually "hollow core," consisting of a light wood outer frame covered by two thin sheets of masonite or fiberboard, separated by cardboard spacers. Any firearm made can easily deliver effective fire right through these doors and into the body of an officer who's careless enough to be standing in front of one of them while knocking. Even a solid wood door can be penetrated by higher-powered fire. *Do not stand directly in front of a door.*

If the structure permits, stand well to the doorknob side of the door, using your baton to knock on the door. Be prepared to draw your sidearm, or have it drawn and pointed downward (away from your feet), depending on the facts confronting you.

At houses with double entry doors, stay well to the side of *both* doors. If the entry door is recessed, use your baton to knock from your place of cover outside the recess, or if the structure is made of wood, knock on the wall, instead of the door.

No matter what kind of call you're answering, do not stand in this dangerous position.

Always stand to the side when knocking at a door. If you have reason to expect hostility, keep well back from the door by using your baton to knock.

When the door opens, look for the hands of the person who's standing there. If you can't see them, ask the person to bring them out slowly where you can see them. ("You don't have anything in your hands, do you, Ma'am? Could you bring them out kind of slow, where I can see? Thank you, Ma'am.")

Then ask where the person is who called (usually, the same person who answers the door), and ask where the person is who's causing the problem. If the troublemaker is visible to you through the front door and doesn't appear to be armed, secure your weapon before moving toward the open door (unless there's some other reason to keep it out, such as in the case of Donald Leroy Adams).

If the troublemaker is not visible to you, have the person at the door call out and try to get him into the room where you can see his hands before you enter. That's better than walking into the front room and having him step around the corner with a leveled shotgun.

Inside. You and your partner should both enter, and *immediately separate to the sides,* so that a person who suddenly enters the room with a shotgun, or unexpectedly pulls a pistol from beneath a sofa pillow, has to make an election on choice of targets (which will slow him down slightly), and so the officer who is not fired on can take him out. Do not stand shoulder-to-shoulder or one behind the other when you get inside the house. As always, don't let anyone get behind you, or in between you and your partner.

214/THE OFFICER SURVIVAL MANUAL

Kitchens are dangerous places for interviews. Not only is the officer in this dramatization exposing his gun side to an upset woman, he's allowing her within arm's reach of knives, glass items, and a hot skillet.

Standing near a plate glass door or window not only makes a target for outside gunmen, it also creates a risk that an angry resident might shove you through it. You're safer keeping a solid wall to your back.

Find out who else is in the house and try to get everyone except small children to a place where they don't pose a big problem (seated at an empty table, for instance). Then separate the disputants, you taking one and your partner the other. Keep them away from areas where weapons are accessible, such as nightstand drawers, desk drawers, and closets. Don't go into the kitchen, where knives, ice picks, iron skillets and hot coffee might be used against you.

While interviewing the disputants, keep your back to a wall—not to a door or open area. Do not stand near plate glass windows, through which you could be pushed. Do not stand on stairways or balconies or near railings. Do not stand in open areas of the ground floor where you're exposed to anyone on the second floor (such as in a foyer or vaulted room). Do not stand near split-level points (including high porches) where you can be pushed off.

Often, the best places for your interviews will be hallways, patios, porches, nurseries, or children's rooms. Don't use an area that hasn't been cleared. Remember to protect your gun side and to keep your distance throughout your contact.

After your interviews, you and your partner will normally switch subjects and cross-interview, then compare stories. When you do this, don't assume that you won't be attacked just because your partner wasn't attacked—there's no predicting the reactions of angry husbands and wives. Maintain your personal security.

Following resolution of the conflict (whether by an arrest, conciliation and counselling referral, voluntary departure of one spouse, or other remedy), keep your eyes on the disputants. If you arrest one spouse, be prepared for the other to turn on you. If possible, try to move the intended arrestee outside onto the porch before making it obvious that you're going to take him into custody. Once the arrestee has been cuffed and searched, get him out of the area as quickly as possible.

Leaving the Scene. Whether or not an arrest takes place, use the same precautions in leaving as you used in your approach. Don't go strolling across the front yard thinking you just worked a miracle of marriage counselling, and exposing your back to a man or woman who may have been playing along with your suggestions just to get you out the door with your guard down. Watch your rear, move quickly at an angle, and be aware of your closest cover.

Whether or not an arrest takes place, don't sit in your parked car down the street or around the corner while you fill out your report or log the call. Get out of the neighborhood before you stop. Otherwise, the angry family disputant may come down the alley with his deer rifle and catch you so intent on your clipboard that it's the last thing you ever see.

Don't drive past the house on your way out of the area. The guy with the deer rifle may be out in his driveway just waiting for you to cruise past. Instead, back up, make a U-turn, or make a right-angle turn to leave the area.

If you're nearing the end of your shift and suspect there might be more trouble from the disputants, be sure to brief the oncoming watch commander and dispatcher on the situation for the benefit of later-assigned officers.

Other Disturbance Calls. The risk-reduction tactics involved in the domestic disturbance call are generally applicable in other types of disturbances. On calls of *shots fired* or *man with a gun,* don't go charging into the area looking for a confrontation. Put troops all around the area to seal it off, contain and isolate the suspect, and use disabling weapons (tear gas, tasers, etc.) to neutralize the suspect. If you're being fired on, or if other facts make deadly force necessary, use your department's firepower capabilities to overcome the suspect.

The most important thing to remember in dealing with snipers and barricaded suspects is to avoid exposing yourself unnecessarily to their fire. Since an armed man can normally only watch in one direction at a time, you can usually create diversions in front of him while tactical teams move against his position from covered and concealed routes.

Where immediate offensive action is not deemed necessary to save lives, your low-risk course of action may simply be to cut off the suspect's utilities, back off, and wait him out.

One of the most dangerous times in a sniper or barricade operation is during lulls in the shooting. Movement during a lull is dangerous, because an officer may conclude that the suspect is finished shooting, or out of ammo, or injured or distracted, and the officer may move unnecessarily, or recklessly, and be gunned down by an attentive sniper who was just waiting for a better shot. Don't move if you don't have to. If you do move during a lull, move as if you were under fire: keep low, zig zag, move quickly, and use cover and concealment.

To get over low walls which are exposed to fire, you want to use the quickest possible method. Military manuals and some officer survival publications recommend the low-profile "roll over" technique. This method has you hug the top of the wall, pull yourself up lengthwise onto the top, and roll over onto the other side. The advantage of this technique is that it presents a low silhouette and attracts less attention.

The disadvantage of the "roll over" is that it puts the entire length of your body up as a target, for too long a time. Instead, I recommend the "hop over." Crouch behind the wall, put both hands on top, and spring over with both feet at once, never stopping on top of the wall (as you're practically forced to do in the "roll over"), but quickly vaulting over and dropping down low onto the other side, and immediately moving off in a crouched run for the nearest cover. Although your silhouette will be a little higher as you cross the wall, you'll be moving so much faster that you present a more difficult target than with a low, slow "roll over." Watch a buddy demonstrate both techniques in practice, and you'll see which way he would be harder to hit.

SURVIVING FIELD THREATS/219

The "roll-over" technique is lower—but slower—and leaves you up on the wall as a target.

The "hop-over" technique makes you a bigger target, but it keeps you in motion and gets you over and off the wall much quicker.

DERANGED PERSONS

Mentally deranged persons kill an average of 30 officers every year. An *armed,* mentally deranged person should be treated like any other armed opponent. Some fearless officers think they can talk anyone into doing anything, and in the process of trying, they may needlessly expose themselves just to show their "good faith," and to convince the suspect he has nothing to be afraid of.

No matter how good you are at persuading *rational* people to do things your way, it is a bad mistake to even consider trying to rationalize things with an *irrational* person. His mind isn't working right. He doesn't respond to logic. He doesn't respond to sympathy. His mind is haywire, and he can't be "talked into" being normal. Don't try it. Especially, don't try it while you're exposed to his aim. All he has to do is hear an airplane overhead, or see a reflection across the street, or suffer a mental delusion that you've turned into a cobra, and he lets loose.

Get it through *your* head that *his* head is a time bomb. You wouldn't try to talk a time bomb out of exploding; don't rely on your ability to talk a deranged person out of shooting you. Isolate, contain and neutralize him, like any other armed suspect.

If the deranged person is unarmed when you encounter him, don't treat him like a helpless child or a drunk. Treat him like a time bomb with a 4-foot radius. Maintain your distance. Don't let him get near your sidearm. Don't underestimate his strength, nor overestimate his cooperativeness. He may be perfectly docile and cooperative one second, and go for your jugular vein the next. Handle with extreme caution.

AMBUSHES

Ambushes have killed 104 officers over the last 10 years. The 12 ambush deaths included in the 93, represented 13% of all officer homicides. For many officers, the thought of being ambushed is more chilling than any other, because an ambush can happen anytime, anyplace, without warning. That makes it the most difficult threat to defend against.

The best defense against an ambush is to make it difficult for someone to plan and carry out an ambush against you. If someone makes up his mind to kill you, he can try it one of 2 ways: he can come to a place where you're easily found, or he can entrap you into coming to him. To thwart the ambusher in the first kind of case, you need to make it hard for him to predict your actions. To guard against entrapment, you need to be alert to tell-tale signs that something's wrong. And in both cases, you must be prepared to take immediate action if you do come under ambush attack.

Avoiding Predictability. An ambusher who is going to seek you out can find you around the station, in the parking lot, around the courthouse, at the donut shop, at the restaurant, at your favorite parking place to write reports or work traffic, at places where you make regular calls, and at places along your regular route.

Your defense against this ambusher is to become less regular in your habits and patterns.

Inside the station, avoid standing around next to windows and glass doors. Be particularly careful about nighttime silhouettes on stationhouse windows. Don't stand in the doorway talking to fellow officers or citizens. Don't stand around on the steps.

When you leave the station and go into the parking lot or structure, scan nearby rooftops and windows. Watch for people on foot in the area walking slowly or lingering about. Don't turn your back on them. Don't stand around in the parking lot talking—get in your car and drive away. As you drive out, watch for people off to your sides. Look for standing vehicles with their engines running, and avoid driving by them if you can.

Use side or rear doors into the courthouse some of the time. Again, scan rooftops and watch for suspicious pedestrians or vehicles as you enter or leave. If you want to discuss cases with fellow officers or attorneys, do so inside the building. Don't stand around outside.

Don't develop patterns of going to the same place at the same time everyday for your coffee breaks. Spread your business around. At outdoor places, stay between your car and the building, where you have some cover against fire from passing vehicles. Don't get so engrossed in telling your partner about your love life that you aren't aware of vehicles and foot traffic around you. Scan rooftops occasionally, and don't stand in one place too long.

Don't make a habit of going to the same restaurant for lunch on any given day of the week or hour of the day. Don't always approach it from the same direction. Don't always park in the same spot. Don't take a booth next to a window if you can avoid it. When you leave, don't stand in the doorway or linger in the parking lot.

Don't pick out a few favorite places in town to sit and write reports or watch for traffic violations. Going to the same shady spot on hot days, or to the same well-lit place at night to write reports makes you so predictable that anyone who wants you has only to show up at your favorite spot and wait for you to come along, like a pig to the slaughter (the pun is intentional).

Don't develop predictable regularity in the way you patrol your beat or make security checks. As soon as an ambusher sees you do the same thing 2 times in a row, he can simply set up and wait for you to repeat your pattern.

Vary your routes. Vary your schedules. Vary your direction of approach. Vary your breaks and eating patterns. If *you* can predict where you're going to be at a given time everyday or night, the *ambusher* can, too. You wouldn't type up a schedule of your planned activities and send it out to known criminals in the area so they could plan an ambush for you; don't accomplish the same result by falling into regular, routine, predictable, careless habits.

Avoiding Entrapment. If the ambusher wants to get you to come to him, he has to attract your attention somehow and channel you into his trap. One way for him to do that is to commit a petty offense in your presence and then lead you into the kill zone. When you see that a person is not trying to hide his offense from you, or is even challenging you to come and get him (such as with reckless driving right in front of you, or throwing a beer bottle out of his car, or calling you dirty names), don't take the bait.

Act as if you intend to follow him, and suddenly turn at right angles and disappear from his rear view mirror. If the street configuration permits, you may be able to follow a parallel route and see where he goes. Keep him under surveillance until you can catch him by surprise at a place of your choosing. If this isn't possible, just go on your way; don't think of this as letting a wise guy get away with something right under your nose—think of it as you refusing to fall into a sucker trap.

Another way for the ambusher to get you under the gun is with a phony service call. You get the call, say to yourself: "Just another routine loud party," and drive right into the kill zone without a second thought.

Once again, the dispatcher or desk officer is the first guardian of your survival. If the caller refuses to give his name, address or call-back number, the dispatcher should pass this information on to you, and you should become immediately suspicious of possible entrapment.

If the nature, time or location of the call is out of the ordinary, this may be a tip-off that you're being set up. For instance, if someone reports a burglar alarm going off at a commercial establishment in the middle of a business day, you should suspect trouble. Likewise, a call to a remote residential area at 2 in the morning to report a stolen CB should arouse your suspicions, since this kind of crime would normally be discovered after sunrise.

With every single dispatched call you receive, you should ask yourself: "Is there anything unusual about getting this kind of call, at this location, at this hour?" If the answer is "yes," have the dispatcher call back and confirm. Approach from the least-likely direction. At night, consider cutting your lights as you go in to survey the scene. And have a back-up unit with you. Be ready to move out quickly at the first hint of trouble.

As you approach the location of any call, look for signs that something's out of the ordinary. If the call is on a loud party, for instance, you should be able to hear the noise from a block away. You should see guests' cars parked in the driveway and along the street near the party address. There should be lights on inside the party house. If these normal indicators are missing, suspect a set-up and pull back.

If the call is a reported traffic accident, you should see people standing in the street and on the curb, pointing and running back and forth. Neighbors should be looking out their doors and windows, or standing in their front yards. A faked accident which lacks screeching tires, banging collisions, and breaking glass won't attract neighborhood attention.

If the call is a crime report to be taken at a residence, the lights should be on (at night), and there should probably be a car parked in the driveway. If the lights are out and the driveway is empty, suspect a set-up and back off.

A person who reports a prowler, a nighttime rape, or a recently-discovered nighttime burglary will usually have the entire house lit up, including all outside lights, and there will be a dog barking. If you see a quiet, dark house, or maybe just the front porch light on, pull back.

Other things out of the ordinary might be the street lights shot out in the block, trash cans set out in the road, forcing you to slow down and go around them, or a "stalled" car out in the traffic lane to channelize you. Maybe there should be pedestrians in the area, and there are none. Maybe there should be vehicular traffic, and there is none. Maybe there's a briefcase lying on the sidewalk, or a lit flashlight lying on the ground, just begging for a sucker to come over and pick it up.

Every time you go into an area on a call, ask yourself: "Is there something out of the ordinary here?" If the answer is "yes," change directions immediately. Pull back and have the dispatcher phone the call-back number for confirmation. Wait for a back-up. Go in from the least-obvious approach route (use an alley or vacant field, for example). If there's only one way in, go in with 2 cars abreast; a show of strength may deter a would-be ambusher.

An ambusher can't possibly know as much as you do about what makes for a legitimate appearance on a call. He has to leave holes in his set-up. If you know your territory, know what to expect, question the normalcy of the call and the scene which greets you, and do the unexpected when you smell trouble, you stand a very good chance of thwarting the ambusher's plans for you.

The one technique which is easiest for the ambusher to use to draw you to an area is to shake the door of a closed business and set off the burglar alarm. This way, the anbusher has nothing to stage, doesn't have to call the department and get his voice recorded, and doesn't risk your spotting unusual conditions that might put you off. All he has to do is shake the back door, climb on top of this or any nearby building, and pick you off as you walk up to check the door.

Whenever you answer a burglary alarm call, remember the possibility that it's a set-up. Scan all rooftops frequently as you go through the procedures discussed earlier. Be watchful for pedestrians, slow-moving vehicles, and parked vehicles in the area. Look for ladders, ropes and other means of scaling nearby buildings. Don't stand around under building lights. Move inside shadows, and be alert to cover options. Don't let your guard down until you're completely away from the area.

Ambush Reaction. If you do come under ambush attack, you have 3 basic options for reacting: you can immediately assault the ambusher's position, you can seek cover and return fire while awaiting help, or you can get out of the fire zone by the most expeditious means.

Direct assault is generally the least desirable reaction. The ambusher has all the advantages, including cover and concealment, often a higher vantage point, often a more powerful weapon with telescopic sights and longer range, and a planned escape route or an intention to commit suicide. You have the disadvantage of being caught by surprise—perhaps out in the open—and not knowing precisely where the ambusher is holed up. Direct assault should be used, therefore, *only* when you can't get out of the area (car disabled and on fire, for example) and there's no safe place for you to seek cover and engage the ambusher while awaiting help (you're in the middle of a level parking lot being fired on from the nearest building).

If you decide on a direct assault, <u>use all the firepower you have</u>. Your shotgun is best. If you don't have a shotgun, pull your back-up weapon and go into the assault "with both guns blazing." Don't stand still. Don't walk. Don't move in straight lines. Crouch down low, run fast, zig zag, put maximum fire into the ambusher's general vicinity (as close as you can bracket his location), and yell like a crazy man to try to unnerve him. In a direct assault, the only chances you have are to get him in a burst of fire, or to scare him into retreat with your audacity. Don't stop charging until you get to a place of cover, or overrun his position.

If you get into a direct assault situation while driving, head straight for the ambusher and give it the gas. Don't slow down and don't stop. Get down low in your seat, hold the steering wheel on a straight course, and plow right through the ambusher's position. Pretend you're driving a Sherman tank, and run into his car, the wall of his house, or whatever else he's using for cover. As soon as you come to a stop, get your seat belt off and your gun in your hand. If there's anything left of your ambusher, don't give him the next shot.

Don't use a direct assault if it's safe to seek cover and return fire. Get to a place where he can't deliver either trajectory or ricochet fire into your position, and sit tight. Unlike the direct assault, you do not want to pump all your ammo into his position. Fire only when you have a target, and don't use up ammo before help can arrive, just because you want to make noise. If you shoot dry, the guy can walk up and shoot you at point-blank range. Just hold in place until help comes.

The best reaction of all, whenever it's possible, is to get yourself out of the fire zone as fast as you can. If you come under attack while driving, make a left or right turn and accelerate away rapidly. Stay low and well back in your seat.

If right-angle turns aren't possible, stop fast, get into reverse, and accelerate backward rapidly, keeping low in the seat. Or, if the street is wide enough, brake sharply, whip a U-turn, and accelerate away, again keeping your head low.

The only time you should simply accelerate straight forward is when there's no place to go left or right, and a gunman is firing from a position to your rear or even with you. In those cases, rock your car from side to side with *short* turning movements of the steering wheel, and turn your car into a dragster getting down the road. Keep low.

If incoming fire or traffic conditions make it impossible to maneuver your car quickly out of the fire zone, drive up onto a lawn or amid parked cars in a lot or up to the corner of a building. Bail out of the car on the side opposite the ambusher, and take cover behind the engine or front wheels until you plan a foot retreat. Dash between points of cover to get out of the line of fire, and then continue your retreat to a safe place.

After departing the fire zone, get troops to come in and surround the ambusher's location. Handle as a barricaded sniper.

The key to successful ambush reaction is to **move immediately and swiftly.** The ambusher only opens fire when he has you in just the right spot. So whatever you do, *don't stay in that spot.* Whether you decide to assault, or move to cover, or beat a retreat, do it immediately and quickly.

PLAINCLOTHES AND OFF-DUTY RISKS

Of the 93 officers recently killed, 61 were assigned to uniformed patrol duty, 21 were detectives, and 11 were engaged in police action while off duty: a full third of the 93 slain officers were not wearing a uniform. Considering the relatively small proportion of officers who are assigned to detective work, and the relative infrequency of off-duty action as compared to on-duty activity, it's obvious that non-uniformed police activity is *proportionately* more likely to prove fatal for an officer than uniformed patrol.

We all know that uniformed officers spend more time in the field, answer more calls, stop more cars, investigate more suspicious persons, respond to more silent alarms, and spend more time doing night work, when the dangers are highest. So why is it proportionately more deadly to take police action while *out* of a uniform, rather than *in* one? I can think of several possible answers.

Risk Factors. One obvious factor is the difference in equipment. The uniformed patrolman is specifically equipped to handle danger: he has a 4" or 6" barrel on his handgun; the plainclothes officer (detective or off-duty officer) usually carries a 2". The patrolman's weapon is more likely to hold at least 6 rounds; the plainclothes weapon probably holds only 5. The patrolman usually carries extra ammo for 3 reloadings; the plainclothes officer carries enough to reload once, if at all. Patrolmen often use speed loaders; the plainclothes officer keeps extra ammo in belt loops.

The patrol officer is more likely to have access to a shotgun when he needs it; he's more likely to qualify with his weapons more often; and he's equipped with a variety of weapons to handle varied threats, including a baton, a sap, a mace sprayer, a heavy flashlight, hobbling devices and extra cuffs. He always has a radio in his car, and often a portable one on his belt, giving him superior communications capability. He has the psychological advantage of his uniform and badge. He's more likely to have the protection of body armor and ballistic devices. He's more likely to have a partner or back-up officer available for help. And he's more likely to be in better physical condition than a detective.

In short, a uniformed officer is prepared to do battle—both physically and psychologically—and he's much better equipped for the job. Those advantages give him a higher level of survival fitness than he has while off duty, or while working as a detective or other plainclothes officer.

Another factor may be differences in training, briefing and supervision. Training for patrol officers is usually ongoing, whereas it's often assumed that detectives are beyond the need for training, because of their prior field experience. Patrol officers receive *daily* briefings on criminal activity in the field, whereas detectives may have only a *weekly* briefing, or none at all. Patrol supervisors are likely to be out in the field, correcting improper work on the spot, while detective supervisors are more likely to supervise from behind a desk, leaving their officers to work pretty much on their own.

Differences in perception of danger and the ability to survive it may also account for proportionately higher detective deaths. Because officers assigned to investigative work are likely to have more years service under their belts, they may be more likely to become complacent, taking careless risks because they've come to believe the greatest danger is behind them, or because they suffer more from a macho-Superman-John Wayne-Clint Eastwood syndrome.

Detective work is more likely to involve a higher degree of contact with more hardened criminals. Whereas the patrol officer spends 85% of his time handling service calls, traffic matters, and drunk drivers, the investigator spends most of his time tracking and arresting burglars, rapists, robbers, dope dealers, and murderers. These are criminals who have more to lose by being captured than do most of the people typically encountered by the uniformed patrol officer.

An officer in plainclothes is difficult to identify as an officer, especially if he's been working vice or narcotics, or has let himself get completely out of shape. Armed citizens, intent on protecting their homes and businesses, don't expect a detective to look scroungy and dirty, to be dressed in old jeans and Budweiser T-shirts, nor to be carrying around a belly that belongs in a Santa Claus suit. When they see such a person with a drawn gun, frightened citizens are likely (and understandably) to draw the wrong conclusion.

The same goes for officer identification by uniformed personnel who are not personally acquainted with every detective or off-duty officer who may be observed in the midst of police action. Tragedies are regularly reported in which plainclothes officers are shot and killed by uniformed officers under mistaken, but justified, assumptions.

Risk Reduction. If these various contributing factors to plainclothes death are obvious, the solutions are just as obvious: an *off-duty* officer can improve his survival chances by wearing his soft body armor, by carrying extra ammo, by carrying his handcuffs, by getting to a telephone and calling for help, by remaining just as alert to danger and cover options as he would if in uniform, by taking no unnecessary risks, and by facilitating officer recognition (I'll discuss ways to do this in a minute).

The *detective* can improve his survival fitness by maintaining firearms proficiency, by carrying cuffs and extra ammo; by wearing soft body armor, by keeping himself informed on street activity, by keeping up-to-date in training, by refusing to submit to the macho-sucker attitude that he's seen it all and knows it all, by bringing in uniformed troops on risky operations, by facilitating officer recognition, and by keeping his body fit.

Detective bureaus can't operate like a patrol division. The nature of the work dictates some differences. But there usually is *no* legitimate reason for a detective to fail to qualify regularly, or to keep in shape, or to keep abreast of training developments, or to use body armor when his outer wear permits, or to observe reasonable precautions for his own safety during risky operations. These are matters over which the detective himself has personal control, and for which he has personal responsibility.

Officer Recognition. A citizen or an officer has 2 means of identifying you as an officer when you're in plainclothes (whether off-duty or investigator): one means is through what he <u>hears,</u> and the other is through what he <u>sees.</u>

The first thing you want to do when you're in a challenged position is to identify yourself *vocally*— not be reaching for your shield or ID card. And when you tell someone you're a law officer, sound like it— don't just say: "Police Officer;" say: "Investigator Morris, 43d Detectives. I'm on the job here," or "Deputy Carter, El Paso Sheriff's Office, off duty."

If you're confronted by a uniformed officer, say something to him that only a cop would know, such as the criminal code section of the case you're on, or some appropriate piece of police jargon: "I'm working a 459—it's Code-4."

Whenever possible, you should pin your badge in the customary place on your civilian clothes when you're going into action, or carry it in your non-gun hand for quick display. On raids and combined operations (detectives and patrol working together), you should wear something distinctive, such as your Sam Browne belt, a raid jacket with insignia stenciled on, a colored plastic or nylon vest with insignia, or an armband or neckerchief designated by your department for plainclothes use.

If you normally wear a suit, a sport coat, or a windbreaker or other jacket while in plainclothes to conceal your weapon, place your badge in such a way that it becomes visible before your gun does when your jacket

SURVIVING FIELD THREATS/235

An armed plainclothes officer is in a precarious position when confronted by a uniformed officer, or an armed citizen.

With your badge in this position, a person who sees your gun cannot help seeing your badge, too. That can remove tension and danger for both of you.

is opened, or when you raise your hands. For example, if you wear your holster on your right side, wear your badge on your belt, just in front of your holster *(not* on the opposite side).

Remember to move slowly when an armed citizen or uniformed officer has you under suspicion. Ask permission to make any move, explain what you want to do and where you have to reach, and use your left hand to reach in for your badge, ID card or weapon.

ARREST AND CONTROL

Arresting, transporting, and handling prisoners has consistently been the single most dangerous category of hostile field activity, accounting for more than 1 of every 4 officer murders over the years. This means that, on an average, approximately 30 officer fatalities occur each year as a result of arrest and control activities.

Once it becomes obvious to a criminal that you intend to arrest him, his greatest incentive for violence arises. Once you move in close to take custody, his best opportunity for attack develops. To protect against the risks involved in arrest and control activities, you need to use procedures which put the suspect at increasingly greater disadvantage and disability, the closer you must get to him.

Disarming. When you have a suspect at gunpoint who has a knife or gun in his hand, keep your shotgun or sidearm aimed at him, remain behind cover (if available), and tell him very firmly not to make any moves until you tell him to. Because a cocked firearm may discharge when dropped or tossed onto a hard surface and because you don't want to put the suspect into position to throw his knife, your best approach is to order him to keep the weapon out at arm's length, pointed to the side (not toward you or anyone else), and slowly bend down and lay the weapon on the ground.

Once the weapon is out of his hand, order him to straighten up with his hands stretched into the air straight above his head. Now tell him to take small steps away from the weapon (designate backward or to his non-weapon side) until you tell him to stop. March him in this fashion until he's at least 10 or 12 feet from the weapon.

If the armed suspect doesn't comply with your orders, hold him at bay until back-up officers can employ non-lethal weapons to incapacitate and disarm him. If these weapons are not available to you, or if the suspect is threatening the public safety with his weapon, it may be necessary to have a marksman shoot him in a non-vital area, such as a shoulder, to neutralize the threat. If no marksman is available, you may have to shoot him yourself (use your pistol, not your shotgun, if your intent is to inflict an incapacitating injury, but not to kill).

If the suspect has no weapon in his hands when confronted, order him to raise his hands all the way up and turn in a full circle, so you can look for visible weapons or noticeable bulges. If you see any weapons, do not tell the suspect to get rid of them—you don't want his hands anywhere near a weapon. You want to get him disabled and then remove the weapons yourself.

Disabling. To disable the suspect without having to get dangerously close, keep your position of cover, keep your weapon trained on him, and order him to follow your directions.

If he's dressed in such a way that a belt is visible, give him the following orders:

"Do exactly what I tell you! Don't make one wrong move! Keep both of your arms straight up in the sky! Turn around facing away from me!"

"Keep your right hand exactly where it is! Slowly, bring your left hand straight out to your left side! Don't bring it in close to your body! Turn your left palm facing backward, thumb down! Now slowly bring your left hand behind your back, just above your belt! Keep your palm facing out!"

"Suck in your stomach and slide your left hand down through your belt! Don't put it inside your pants—just down through your belt! Push it down as far as it will go!"

"Now, slowly stick your right hand straight out to your right side! Move real slow and careful, and bring your right hand back and put it through your belt! Keep your palm facing me!"

"Right where you're standing, go down slowly to your knees!"

"Pick up your right foot and cross your right ankle over the top of your left ankle!"

"Now walk your knees as wide apart as you can get them! . . . Wider! Don't lean back on your heels—just sit straight up! Now, don't you move another muscle, do you understand me?"

Don't let him lean back, or his hands may be able to reach into his boot top, or into an ankle holster. Be sure his knees are stretched wide apart. In this position, if he tries to uncross his ankles, he stands a very good chance of pitching forward and busting his face against the ground. (Try it and see.)

Hand your shotgun to your partner or back-up officer, and draw your sidearm, holding it in close as you approach the kneeling suspect. If you don't have a partner, unload the shotgun, pocket the shells, and put the shotgun down behind the suspect's feet where you can watch it.

Approach the suspect directly from his rear. Don't allow him to turn his head or move in any way. If you have a partner, he should keep his pistol trained on the suspect from an angle that would allow him to shoot without endangering you.

When you get to the suspect's feet, slide your leading foot underneath his crossed ankles, keeping your gun side away. If he makes any trouble, pull your foot straight up, and he'll fall forward on his face.

Holster your gun and snap your safety strap. Get your cuffs out in your strong hand, and grab his shirt collar with your weak hand. If he's wearing no shirt, place the flat of your hand against his upper back, between the shoulder blades. In either of these positions, you can slam him forward if he makes trouble.

With his hands still through his belt, cuff both wrists, back to back, palms out. His middle belt loop and the width of the cuffs will prevent his pulling his hands back

DISABLING A FELONY SUSPECT

1. Get the suspect's hands into the air.

2. Left hand extended palm rearward.

3. Left hand through belt.

4. Right hand extended palm rearward.

SURVIVING FIELD THREATS/241

5. Right hand through belt.

6. Have suspect kneel in place.

7. Suspect crosses his right ankle over left.

8. Order suspect to "walk" knees wide apart.

242/THE OFFICER SURVIVAL MANUAL

9. Keep your gun side away and hook your opposite foot under the suspect's crossed ankles.

10. Cuff the suspect's hands beneath the belt, on each side of the belt loop, for maximum restriction.

11. Controlling the suspect by the arm, lower his head into contact with the ground. This suspect is disabled and ready for searching.

through his belt. Double lock the cuffs. This suspect is now ready to be searched.

If the suspect has no visible belt when you encounter him, get his hands straight up, talk him into the kneeling position while keeping his hands up, and then proceed as follows:

"Very slowly, point your left hand straight out to your left side, thumb down, palm facing the rear! Keep your right hand straight up! Bend your left arm and slowly bring your left hand behind your back—keep it way up on your back, fingers pointing at me, thumb up! Hold it still!"

"Slowly stick you right hand straight out from your right side! Get your thumb down, palm facing toward the rear! Bend your right arm real slow and bring your right hand behind your back—get it up high, thumb up, fingertips pointing straight at me!"

"Now, don't you move another muscle, do you understand me?"

Make your approach and get your foot beneath his ankles. Keep your gun drawn in the close quarters position. With your weak hand, get a good, firm grip on his nearest thumb, putting just a very slight backward pressure on it—not enough to cause pain, just enough for control. If he causes any trouble, snap his thumb backward forcefully, all the way down to his wrist, breaking his thumb, and then pick your foot up quickly to slam his face into the ground. Since his hands are free, step back and draw down on him.

244/THE OFFICER SURVIVAL MANUAL

1. If the suspect has no visible belt, have him cross his hands high on his back, palms rearward and put him into the kneeling position. As you advance to get your foot beneath the suspect's crossed ankles, keep your weapon in close quarters.

2. Firmly grasp the suspect's thumb and apply slight backward pressure to let the suspect know what will happen if he resists.

3. While you maintain control on the suspect's thumb, holster your gun and snap the safety strap. Cuff the suspect and double-lock. Lower his head to the ground and search.

As soon as your leading foot is in position and you have a fist locked onto his thumb, holster and secure your gun. Remove your cuffs. (In this operation, as in most others, you're better off having your handcuff case situated where you can get to it with your gun hand.) Using your strong hand, cuff his right wrist, and then his left, as discussed earlier. Double lock. The suspect is ready to be searched.

Don't try to cuff and search more than one suspect by yourself. Wait for help. Get them into a prone position, if possible, while you're waiting. To do this, keep their hands all the way up, have them face opposite directions, 2 or 3 feet apart, and then talk them through the kneeling position, ankles crossed, except *don't* spread their knees apart. Have them fall forward, catching themselves on their outstretched hands. Walk their hands forward until they're proned out. (You can position 3, 4 or 5 suspects into a star shape in this manner.)

Have them stretch their arms to the front and turn their palms up (this is less restful than allowing their arms out to their sides, and it brings on muscle fatigue in their arms and shoulders quicker). Keeping their ankles crossed, have them bend their knees and stick their feet up into the air (it would take them longer to get to their feet from this position than if you allow them to leave their legs flat on the ground—try it and see). Order them to keep their chins tucked into their chests, and their foreheads touching the ground. Don't let them talk.

DISABLING MULTIPLE SUSPECTS

1. Suspects face opposite directions, hands in the air, a few feet apart.

2. Order suspects to their knees.

3. Suspects fall forward onto their hands.

4. Order suspects to lie face down, arms stretched forward, palms up. Having the suspects cross their ankles and raise their feet into the air makes it impossible for them to suddenly jump to their feet.

248/THE OFFICER SURVIVAL MANUAL

When help arrives, keep the prone suspects where they are and talk their hands behind them, either through their belts or high on their backs. With your back-up covering, approach at an angle from the rear until you're even with the first suspect's feet (handle the most threatening suspect first). With your weak hand, grab the suspect's rearward foot and put a slight twisting pressure on it, forward and down. Holster your gun. If the suspect moves or causes trouble, twist violently down on his foot, breaking his ankle. Step back and draw down on him.

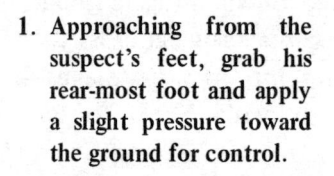

1. Approaching from the suspect's feet, grab his rear-most foot and apply a slight pressure toward the ground for control.

2. Firmly controlling the suspect's foot with your weak hand, cuff him with your strong hand.

If he causes no problems, maintain a firm grip on his foot, and cuff him, as described earlier. Before turning your attention away from this suspect to cuff the others, search into any area where he can still reach for a weapon, such as inside his back waistband, back pockets, boot tops and ankles. This is just a partial, preliminary search, for protection while you're cuffing his accomplices.

You can't always use the prone technique. If you're on a street or sidewalk that's reflecting 100 degrees of summer sunshine, it won't work. If you're in a parking lot that's covered with ice and snow, it won't work. When you have multiple suspects in conditions where the prone position won't work, put them into the kneeling position, facing opposite directions, with their hands high in the air until help arrives. If that's not practical, have them stand facing a wall or other barrier, keeping their hands up. Don't let them talk.

I've often seen the prone position touted as the best to use. When you have only a single suspect, however, you'll disable him more, and achieve greater control, with the *kneeling* position I've described. To find out whether or not you agree, put yourself in the suspect's position with a buddy controlling you; practice both the kneeling and prone positions as suspect, and then switch and practice both as officer. You'll see the difference.

Prone and kneeling positions are used to allow you to disable felony suspects. You generally will not use these positions for *misdemeanor* suspects if you don't have reason to believe they're armed or dangerous.

The traditional technique for handling misdemeanor suspects in the field has been to prop them up against a wall or the side of your patrol car, conduct a search, and then cuff them. Because this technique led to officers losing their weapons, some authorities are now recommending against the prop technique in favor of a standing search, conducted while the standing suspect is kept off balance with his back arched back. Proponents of the standing search suggest you hold onto the suspect's interlaced fingers behind his head, while searching his entire body with your strong hand.

I don't particularly like either one of these approaches. If you're a 5'6" cop, trying to keep one hand behind the head of a 6'2" suspect while searching over the rest of his body with your strong hand, you're just in a ridiculous position, trying to do an impossible task. A person's arm span is the same as his height. A 5'6" cop can only reach 5'6". With one hand holding a tall suspect's hands behind his head, the 5'6" cop simply cannot reach over to the suspect's opposite boot top to look for weapons.

I've experimented with the standing search technique with officers and "suspects" of varying heights and degrees of strength, and I think it's an awkward, dangerous technique for most officers. What's more, this technique suffers from the same basic problem as the traditional prop method: it's *backward*. It puts the search *before* the disabling; it forces you to get into positions which compromise your security, before you've done anything to effectively disable the suspect.

As you noticed, I labeled this topic "Disabling." We haven't gotten to searching yet (it's the next topic). And there's a very good reason for this order of discussion: you unnecessarily reduce your survival superiority when you undertake to search someone *before* he's disabled and cuffed. That's doing things backward. The better approach is to disable and cuff, and *then* search.

An uncuffed suspect can cause all kinds of problems while you're bending down all around him like a one-armed masseur: he can kick you; he can bring an arm down and pound your head; he can fall on you; he can pull one arm away and get to a concealed weapon; or he can catch you just at the right angle and get his hands on your gun. The problem with the prop search never was in the *position*—it was in the search-and-disable *sequence*. The answer wasn't to devise an awkward standing search that retains this dangerous, backwards sequence—the answer is to change the sequence from search-and-disable to *disable-and-search*. Once you have a suspect disabled and cuffed, you can control him and search him in much greater safety.

Many academies (including the one I went to) have taught officers the search-and-disable sequence for misdemeanor arrestees. It's widely used throughout the law enforcement field. That doesn't make it right, of course. What's right is what works, and works well. For hundreds of officers who get themselves assaulted or killed while in the process of searching uncuffed suspects, this backwards sequence doesn't work very well.

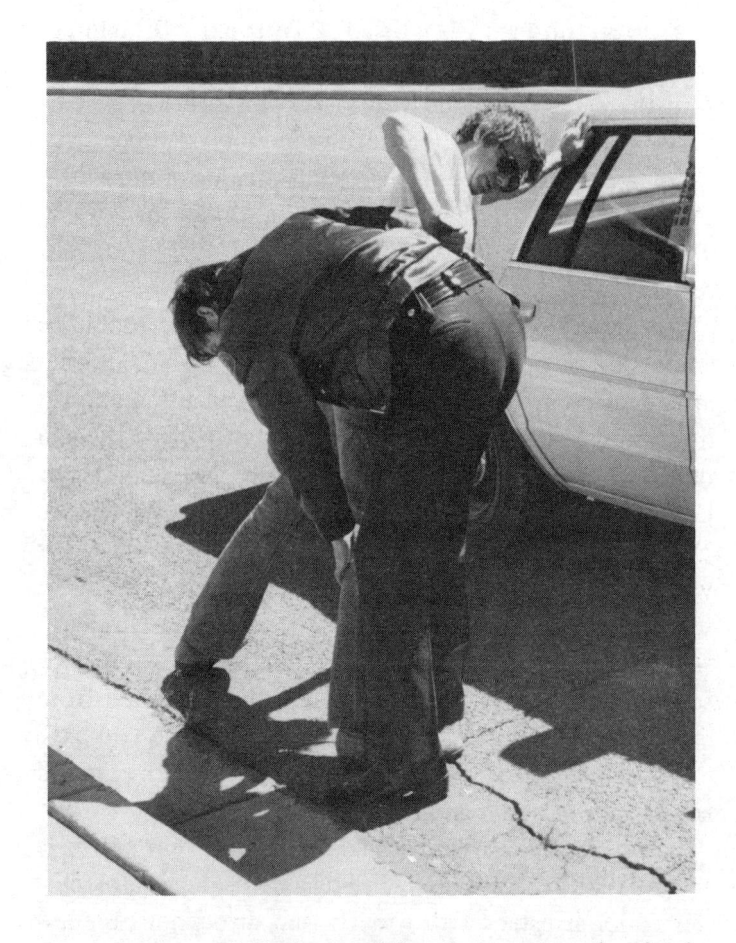

The problem with the traditional prop search of misdemeanor arrestees was not the position, but the dangerous, backward sequence of trying to search an *uncuffed* suspect.

My recommendation for disabling the misdemeanor arrestee and getting him ready to be searched is that you order him into the prop position against the wall or your car, while you maintain your 4' safety interval and protect your gun side. Have him lower his head into contact with the wall, and then walk his feet backward until he's at about a 30 degree angle with the wall. Keep his feet well apart.

Next, order him to hold his weight up with his head and bring his hands behind him, either sticking them through his visible belt, or holding them high up on his back, as described earlier. With cuffs in your weak hand, approach directly from his rear. Standing between his legs with your gun side to the rear, hook your leading foot just inside his opposite foot. Take the cuffs with your strong hand, and place your weak hand against his back, between the shoulders. Cuff as already described. Double lock the cuffs. This suspect is ready for searching.

Once you have the misdemeanor arrestee in the 3-point stance (head and feet supporting him), with his feet well back, he can't push off; he can't kick. And once you have him cuffed, he can't draw your weapon during a search; he can't hit you in the head; and he's much more limited in his ability to draw a concealed weapon. With your foot hooked inside his, you can pull his foot out from under him if he causes trouble during the cuffing, and send him crashing into the ground.

A propped suspect in this position can be approached and cuffed without the risks of the technique illustrated on page 252.

Using this approach, you can search the cuffed suspect in the prop position while restricting the suspect's opportunities for resistance.

Searching. To search the cuffed and disabled *kneeling* suspect, hold onto his arm and lower him forward until his forehead touches the ground. Have him uncross his ankles and spread his feet apart. As with any suspect, you first want to search any area that's accessible to his cuffed hands. Feel inside his back waistband, search his back pockets, search jacket side pockets, and inside shirt and jacket cuffs.

Next, begin a systematic search from head to toe on his right side, and then his left, including his long hair, headbands, neck, collar, shirt sleeves, armpits, sides, back, chest, stomach, hips, abdomen, groin, thighs, behind the knees, legs, inside boot tops and sock tops, and shoe bottoms (razor blades, coiled wire, and small knives and saw blades may be taped to shoe soles).

If the felony arrestee is a woman, check below and between her breasts. Through her outer clothing, run your fingers along the outline of her bra strap across her back, feeling for suspicious bulges. If you saw her stuff something into her bosom during your approach, or if you have some other reason to suspect that a dangerous weapon may be concealed inside her bra, loosen the straps and shake the bra. If a female officer is on the scene, she should conduct the search of a female suspect, because she's more likely to conduct a thorough search (male officers may be reluctant to be thorough because of the fear of charges of impropriety).

Check slings, casts, bandages, band-aids and wide bracelets, beneath which criminals sometimes conceal small weapons.

Remove anything you find which could be used in any manner as a weapon, including pencils and pens, jewelry, combs, keys, nail clippers, hair pins, and hair clasps. Examine belt buckles for the kinds of devices illustrated in **"Disguised Weapons,"** later in this chapter.

To search a prone suspect, raise him up into a kneeling position, and then lower his head to the ground and spread his feet apart. Conduct the search as described above.

Search the propped misdemeanor suspect where he stands, keeping your foot hooked inside his on the side you're searching, and maintaining a firm grip on his arm with your nearest hand, for control.

When searching someone following a limited, preliminary search, don't skip over the areas you've already searched. The suspect may have managed to move an item into that area, not expecting you to check it again.

Just because you find a weapon during your search doesn't mean the search is over. If you find *one* weapon, assume he's carrying *two,* and keep looking. If you find *two* weapons, assume he's carrying *three,* and keep looking. The search is over only when you've cleared every possible hiding place on the suspect's body and in his clothing.

If you have PC for a pat-down search (frisk), but not for an arrest, stand to the side of the suspect, keeping your gun side away, and pat him down with your weak hand. Then go to the other side and repeat this process, again keeping your gun side away. Do *not* stand face to face with the suspect and reach both your arms toward

SURVIVING FIELD THREATS/257

Do not put yourself into this risky position during a "frisk," or "pat down" search.

Protecting your gun side, conduct pat down searches from the side.

him. He can take his pick of either breaking your arms, kicking you in the groin, or going for your gun.

If you're going to transport someone who is not under arrest (such as an accident victim or stalled motorist), require them to submit to a pat-down. Such passengers have been known to pull weapons and stab or shoot officers in the back of the head.

Transporting. Traffic permitting, open your right rear door to place your suspect into the right rear passenger seat, furthest away from the driver. If your suspect is cooperative, get him to his feet and order him into the car. Stay to his right rear, and don't get into a position where he can kick the car door against you. Strap him in with the seat belt. He's ready for jail.

If your suspect is uncooperative or is a suspected felon, maintain control over him with an appropriate hold (see **"Use of Force"**) and deposit him into the car. Have your partner come through from the opposite side to help you, if needed. Buckle the suspect in with the seat belt. To keep him from kicking holes in your car, remove his shoes. If you have a hobble, use it. This suspect is ready for jail.

Do not transport an uncuffed prisoner. Even if the prisoner is a passed-out drunk, a cute young lady, or someone's grandma, cuff *every* prisoner you haul. Uncuffed prisoners kill sympathetic officers.

Do not grab hold of the handcuff chain when handling cuffed prisoners. A prisoner who violently twists his wrists across each other can catch your fingers in the links and do serious damage to your hand.

If your 2-man patrol car has no cage, the second officer should sit directly behind the driver, to the left of the suspect. This position gives the officer better observation of the suspect during transportation.

Once you take custody of a suspect, do not allow anyone to get near him. Often, as you're hooking up a suspect, his wife or friend will want to get his keys from his pocket, or give him a kiss. Don't be a nice guy and permit this kind of dangerous contact. If something needs to be exchanged, you be the middleman. Other than that, just tell bystanders they can see the arrestee during visiting hours at the jail.

USE OF FORCE

Obviously, not everyone is going to cooperate with your efforts to arrest, disable, search and transport them in a peaceful manner. Each year, approximately one of every sixteen officers in the US is assaulted—a total of 56,130 assaults against officers are currently reported. The total of more than 56,000 reported assaults typically leaves 90 to 100 officers dead and some 22,000 injured. In the face of an increasing willingness by criminals to use force against you, you must become increasingly adept at avoiding or responding to confrontations involving force.

Avoiding Hostility. Better than being able to *win* a fight is being able to *prevent* one. There's nothing shameful or cowardly about talking your way out of a fight, or stalling an inevitable fight until your reinforcements arrive. It's in the interest of your survival to avoid physical hostilities whenever you can do so without compromising your law enforcement duties.

One way to avoid unnecessary fights is to avoid challenging someone unnecessarily. Don't go into the family disturbance scene or the bar disturbance or the loud party with a chip on your shoulder. Don't be pushy with someone who's surrounded by his friends or family, where his pride is going to force him to push back. If you have to let this guy know he's out of order, try a polite request, rather than a gruff command, to get him to change. Give him a way to comply without his having to lose face in front of his friends and family. If these efforts don't work, ask him to talk to you outside, away from the people he's trying to impress, and then make it clear to him that you're going to enforce the law one way or another, and that he's better off taking the way out that you're giving him.

I once worked with a partner who was a master at getting cooperation from even the biggest, meanest, drunkest men in town. Although this officer was in excellent physical condition and was a martial arts expert, he never came on like he was anxious to prove his superiority. He always offered the other guy a peaceful, face-saving way out, and the other guy usually took it.

For example, we once went into a bar in the roughest neighborhood in town on a "keep the peace" call, and found a guy who looked like the Incredible Hulk creating a disturbance. Instead of running up and grabbing this big guy, my partner just took out his traffic whistle and blew it to get everyone's attention. When he asked the Hulk what the problem was, the Hulk said: "None of your business, Cop! You think I can't whip your ass?"

My partner said: "Oh, I don't suppose I'd even be any contest for you. But I go off duty in an hour and I've got a date with a hot little mama who's expecting me to come over in good shape, and if you do anything to mess me up, she'll come over here and turn you every which way but loose."

The Hulk and everybody else in the bar started laughing. The trouble was over.

A few weeks after that incident, we were trying to break up a fight between half a dozen bikers from two rival gangs in the parking lot of another bar. Suddenly, bodies started flying through the air, and the bikers started scrambling to get away. We looked around, and there stood the Hulk, grinning from ear to ear. He said: "I figured I'd better get these boys off of you before they go and make that little mama of yours mad."

Not everyone will take the easy way out. In fact, some people may mistake your offer as a sign of your weakness or insecurity, and get even bolder with you. When that happens, you want to correct their misimpression right away. The best thing to use for that is your *command presence*.

Command presence is made up of several different things, including the way you hold your head, set your jaw, narrow your eyes, point your finger, and put authority into your voice. It may include things you do to "clear the decks" for action, such as snapping the chin strap on your helmet, putting your hand-held walkie-talkie onto your belt, and gripping the handle of your baton or resting your hand on your gun butt. All of these things carry a message to your challenger that you're ready and able to back up your commands, if the easy way out doesn't work.

If you're not accustomed to taking charge and giving orders, you'd better get used to it, even if you have to practice in the mirror. Street toughs can smell your weakness, your hesitation and your insecurity. They can tell when you're bluffing. They can tell when you're wading in over your head. You want to learn to project confidence and control with your voice—not just with *what* you say, but also with the *tone* you say it in. There's some difference, for example, between "Sit down," and "YOU SIT DOWN!"

You don't have to snarl, and you don't have to yell. But you do have to put unmistakable firmness into your voice. Let your audience know that you're in charge, 100%.

Limited Force. In those situations where hostility simply can't be avoided and you need to use force to make an arrest, you're required to limit your use of force to the minimum amount needed to carry out your duties. In many cases, that means merely using an arm lock or a simple come-along. Effective use of these devices can help keep a confrontation from escalating into a major fight.

SURVIVING FIELD THREATS/263

A low-profile come-along.

Control from elbow lock and wrist pressure.

Slight upward pressure with the baton, while controlling the suspect's collar, makes this a good come-along, even for drunks.

This may look obscene, but it's a very simple, yet effective, come-along for male suspects. The suspect will respond to slight upward pressure by tiptoeing anywhere you tell him to.

When more force is needed, you should prefer using your baton to using your fists: your baton extends your reach while keeping you away from your opponent; it can concentrate more force than your fists; and it can keep you from bruising your knuckles or breaking your hand against your opponent's hard spots.

To make sure your baton is handy when you need it, *always* have it in your ring when you leave your patrol car. Since any kind of call or field contact can turn unexpectedly violent, don't make a habit of leaving your baton behind on any particular kind of activity. Carry it with you.

Baton jab to solar plexus   **Power blow to collar bone**

SURVIVING FIELD THREATS/265

Snap-swing to the elbow.

Snap-swing to the knee.

Power chop to collar bone

Underhand swing to groin

266/THE OFFICER SURVIVAL MANUAL

If you have to resort to bodily weapons, prefer using your feet to using your fists: you can kick further and harder than you can punch, and you're less likely to injure yourself or get grabbed by your opponent if you kick, rather than punch.

You can kick further than your opponent can punch.

Side-kick to the kneecap.

Swing-kick to the shin.

In close-quarters fights where you have to use your hands, don't go swinging wildly, or pounding on places where your blows won't have much effect (such as skull, shoulders, back or chest). Jab into a vulnerable area, where you can do some damage, such as your opponent's ears, eyes, nose, mouth, throat or groin. If you're in danger of losing the fight, it's time to play dirty: pull hair, scratch, bite, spit in his eyes, grab his testicles and squeeze, put a knee in his groin, pinch his neck, twist his ear, grab his thumb or finger and break it backwards—do anything you can to cause him pain and distraction, so you can get back on top. Don't worry if you have to do a few things John Wayne would never do—the main thing is to inflict more pain on your opponent than he inflicts on you.

Deadly Force. Rules vary by jurisdiction on when you're authorized to use deadly force. As I said earlier, you should be sure you have a clear understanding of statutes, departmental policy, and your personal code on use of deadly force. In *all* jurisdictions, you may legally use deadly force when it reasonably appears necessary to protect your own life.

Although it's possible for you to kill someone with your hands, your feet, or your baton, your most likely means of delivering deadly force is with a firearm, since a life-threatening opponent will usually be using a firearm against you (98% of slain officers are shot to death). This makes your familiarity and proficiency with your firearms critically important.

To assure your maximum survival fitness in the event of a gunfight, you can't afford to rely solely on department target range shooting to develop your gun battle skills. Some departments require their officers to qualify with pistols as infrequently as 4 times per year, and with the shotgun only once in a lifetime (at the academy). Some departments use a fixed-distance shooting range, at which you carefully aim at paper targets, in the daylight, and fire only on command. Such practice shooting, with such infrequency, is just about enough to get you killed during a real gunfight.

Most fatal officer shootings take place at night, on the street or highway, at distances of less than 3 yards, consuming less than 5 seconds. They are quick, sudden, close and dark. *That's* the threat you should be training to meet—not the prospect of a paper silhouette popping out at you from 25 yards away and waiting for you to draw a sight picture. Target shooting has only one beneficial use, and that's to promote your familiarity with your weapon. It has nothing whatever to do with teaching you to win a gunfight. Don't depend on it for that.

No matter how often your department requires you to qualify, you should be shooting once a week. Half of this shooting should be done at night. All of it should be done at ranges of 2 to 20 feet.

Your "course" should incorporate police cars, civilian cars and vans, street curbs, trees, utility poles, fire hydrants, building corners, stairways, moving targets, street noises, flashing colored lights and lots of surprises.

"Bullseye shooting, 25 yards. Load 6 rounds. Fire 1 round on each blast of the whistle, single action, point shoulder position."

To illustrate how unrealistic this kind of target range shooting is for the street cop suddenly confronted with a close, armed opponent, I've used the same caption on the next photo. If you think the caption sounds ridiculous with that picture, you're absolutely right.

"Bullseye shooting, 25 yards. Load 6 rounds. Fire 1 round on each blast of the whistle, single action, point shoulder position."

If your department doesn't have such a course, see about setting one up with other agencies on a regional basis. Rotate the "props," the targets, and relative positions each time you shoot the course. Change the light and noise patterns frequently. Don't get used to repetition—no 2 gunfights are ever the same.

Although it's impossible to simulate the stress that comes from knowing someone is shooting at you and trying to kill you, you can achieve a stress-like condition for your combat shooting by exercising vigorously for 5 to 10 minutes immediately before shooting the course. Vigorous exercise will bring about shortness of breath, heavy breathing, increased heartbeat, and muscular tension, similar to several of the symptoms of gunfight stress. What will still be missing, of course, are danger and fear.

If you can't get access to a combat shooting course, go hunting whenever possible. Wear your Sam Browne belt and practice drawing and shooting at moving targets. If nothing else is available, shoot at trees, bushes or dirt embankments as you move for cover (observe hunter safety).

In metropolitan areas where neither combat ranges nor game hunting is available, you may simply have to use the target range for live firing, and simulate combat firing in a safe place where the setting is more realistic than on the target range.

Under any of these possibilities, *frequency* and *variety* are more important than duration. Few officer gunfights are of a sustained duration—in most, only 1 or 2 shots will be fired. Therefore, it's better for you to shoot 6 rounds, 4 times per month, than to shoot 24 rounds at one session. Likewise, it's better for you to shoot 1 round from the shotgun each month than to shoot a dozen once a year.

Most survival instructors have a favorite shooting position to recommend, such as the "Weaver stance," the "isosceles stance," the "Dunbar position," or the "FBI stance." These positions have you place your feet at certain widths and angles, hold your shoulders and arms a certain way, grip and support the pistol with both hands, and aim at your opponent's center of mass, squeezing the trigger with a gentle, uniform motion. *This is all just more classroom nonsense that can get you killed.*

Do you think a criminal is going to take time to go into the Weaver stance before he guns you down? More to the point, do you think he's going to allow *you* time to go through a choreography routine before he guns you down? Not on your life. The criminal never heard of the Weaver stance, and he couldn't care less. While you're foolishly trying to get into some fancy seminar shooting position, the poor, ignorant, untrained criminal is simply going to point his gun and shoot a hole through you.

I keep referring to some of the standard tactics as "classroom" or "seminar" material, because I don't know where else these tactics could have come from. They certainly didn't come from gunfight experience. I would estimate that I've been shot at more than 300 times. I've shot more than 30 people. Not once did

I ever have time to worry about such niceties as stance, grip, aim or trigger squeeze. Not once did I worry about whether my arms were at a 30 to 45-degree angle. Not once did I get a sight picture. Not once did I take evenly-spaced, single shots. Not once did any of these things even cross my mind during a firefight. They're not going to cross *your* mind, either. If they do, nothing else ever will again.

Under fire, you don't have time for nonsense. You don't have time to waste. You either get some lead into the guy who's shooting at you, or you don't have any time left.

I've seen survival books spend 20 and 30 pages telling you how to shoot in a gunfight. I can tell you in 3 words: **point and shoot!** That's all there is to it. That's all you have time for. When you're 3 to 5 seconds away from eternity, you simply don't have time to screw around getting into the isosceles stance. *Point and shoot!*

Notice, I said "point," instead of "aim." You can't aim unless you get your gun up to eyeball level. That takes time. It may take you a full second. And a full second just might be a third of the rest of your life. Don't waste precious time trying to line up your gunsights, your eyeballs, and your opponent's center of mass—just *point* your gun at him and *shoot!*

When you're in a legitimate, survival-threatening shooting situation, don't shoot to scare. Don't shoot to wound. If you fail to neutralize your opponent with your first burst of fire, you're going to give him

SURVIVING FIELD THREATS/275

When you unexpectedly come under fire and don't have time for careful aiming and fancy footwork, just POINT AND SHOOT.

a chance to take more shots at you. And *he* won't shoot to wound.

When I went through police recruit academy, some joker who was teaching officer survival told us that in case of a shoot-out, we should place one well-aimed shot at the opponent, and wait to see if that did the trick before firing another round. This guy was obviously more concerned about lawsuits and community relations than about officer survival. If you're forced to shoot someone, you want to fire a burst of 2 to 4 rounds at him in quick order. Statistically, at least half of your shots will miss. A single round may not neutralize your opponent, and if it doesn't, he's going to be pumping lead into you while you're standing back to assess each round. Don't fool around with

If you have a choice, avoid shooting *over* cover, since this technique exposes too much of your body to hostile fire.

Try to shoot *around* cover, exposing no more of yourself than is necessary to see where you're shooting.

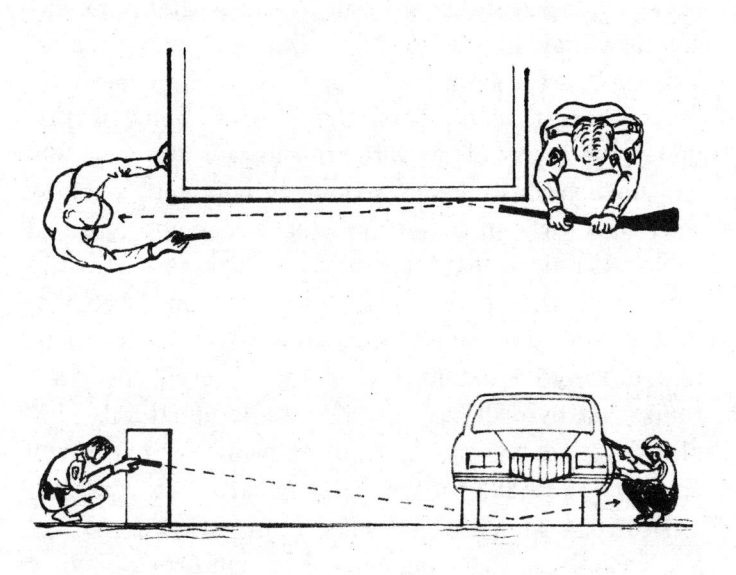

RICOCHET SHOOTING

If you fire against a hard surface at a shallow angle, the lead slug will flatten out and travel along a few inches away from the impacted surface. Although it's possible you might have a chance to use this kind of shooting in a rare situation, as illustrated, you primarily need to be aware that ricochet fire may be used against you (probably by accident) if you don't use cover properly. Don't try to use ricochet fire if you have another choice—it's too unpredictable and uses up your ammo without offering a good probability of neutralizing your adversary.

well-aimed, evenly-spaced, single shots—shoot double-action as rapidly as you can, and put several rounds into your target before you stop to survey the damage.

Another standard classroom survival point is that under the stress of a gunfight, you will automatically and unconsciously revert to trained conduct. Luckily, since most of your training may have been poorly suited to your survival, that classroom maxim isn't necessarily true. Under stress, you may revert to trained behavior *if* it doesn't go against your natural survival instincts; if it does, your natural instincts will prevail. So don't think that by drilling yourself in some unnatural, complicated, time-consuming shooting maneuver that it will come out naturally for you in a gun battle. It won't.

Drilled training which makes use of your natural tendencies—such as crouching when you fire—will come automatically under stress. Drilled training which runs contrary to your natural tendencies—such as lining up a sight picture—will only create a time-consuming conflict for you under stress, slowing down your natural reflexes. To maximize your ability to speed up your natural reflexes, therefore, you don't want to practice unnatural procedures. What you want to practice is what you're going to do instinctively in a firefight: *point and shoot*.

As evidence that you automatically revert to your training in a shoot-out, survival theorists point to an incident in which a California Highway Patrol officer put his empty brass into his pocket during a firefight, as he had been trained to do on the firing range. From this incident, theorists draw the conclusion that the officer was reacting unconsciously, under stress. I doubt that that was actually the case (we'll never know, since the officer didn't survive the battle).

In fact, 3 other CHP officers who were killed in the same battle did not pocket their brass, even though they had the same training. Hundreds of officers in other shootings over the years who had similar training did not pocket their brass. This isolated incident hardly proves a general proposition that you will revert to trained conduct under stress.

The more likely explanation for the fact that this one officer bothered to put empty shell casings into his pocket in the midst of a heavy firefight is either that he didn't want the enemy to see or hear the brass dropping onto the street, thereby revealing that he was in a vulnerable, reload situation, or else the officer was seeking normalcy.

During my first few times under close fire, I found myself doing things that didn't make much sense from a survival standpoint, such as brushing dirt off my clothes, salvaging empty M-16 magazines and .45 clips, and glancing at my watch to see what time it was. (Other soldiers, as I later discovered, did similar, seemingly-illogical things during firefights.)

Now, I was never *trained* to do any of these things. It didn't really matter, in the middle of the jungle, whether or not my uniform got dirty. Magazines and clips weren't going anywhere; they would still be there when the fight was over. And with blood soaking my shirt from a back wound, the man on my right with most of his head blown off, and the man on my left without any face left, his brains running out through his ears, it didn't make any difference what time it was. So why did I do these unnecessary things? I wanted normalcy.

We don't have any experience at facing death. We don't have any training for it. We don't have any prepared mechanisms for coping with the sickening fear of our own imminent death. Our minds can only tolerate so much of that stressful fear and still function.

To counterbalance the unfamiliar feeling of that situation, I did little things to allow my mind to focus for a few seconds on something *familiar*—on something that would only matter under *normal* conditions. To control my fear, I let myself hope that once the fight was over, it would *matter* again that my uniform be clean, that I have magazines to reload, and that I be able to tell how long the fight had lasted. Maybe, with his fellow officer dead and himself outnumbered and outgunned, that CHP officer had to tell himself that it was going to *matter* after the fight that he had all his empty brass saved in his pocket. Maybe his mind needed the relief of a familiar activity.

Mind you, I'm not recommending that you take precious time in a firefight to do normal things for the benefit of your mental stability. You should be using all of your physical and mental resources, and every second of your time, concentrating on surviving and neutralizing your threat. But if you find yourself in a sustained battle doing something illogical, don't jump to the conclusion that it's because of training—it may just be your mind's way of trying to survive the unprecedented fear.

Reloading. Anytime you're involved in a shooting, use every safe opportunity to reload. In a sustained gunfight, you may have a chance to reload after firing 3 or 4 rounds, whereas if you waited until you were empty, there might be no safe opportunity. If you're behind cover and your partner is busy emptying his gun at the crooks, you should be busy reloading yours, so you'll be able to take up the fire while he reloads.

If armed opponents are advancing on you while you're reloading, you may not have time for a full reload. Shoot 2 or 3 rounds at them, if that's all you have ready. If you use a revolver, make sure you know which way the cylinder turns, so you don't line up empty chambers.

Adjustment. In war, where everyone who survived was a killer, there was no adjustment to be made to the act of taking human lives. It was a simple matter of survival, and a commonplace event.

You're in a different situation. Most officers will retire without ever having fired at a human target; few will ever kill in the line of duty. That makes a killing by an officer something of a special event, which is the focus of intense publicity, departmental scrutiny and peer reaction. It is all of this sudden, probing attention which creates most of the stress for an officer who has killed—and *not,* as psychiatrists generally speculate, the realization by the officer that he has killed a human being.

Peers and supervisors can substantially lessen the adjustment stress for an officer involved in a fatal shooting by giving him less to adjust to. The more dramatic and visible the reactions of fellow officers, the more amplified will be the involved officer's perception of his adjustment task. So to assist an officer in getting back to normal after a shooting, other members of the department should avoid blowing the event out of proportion, or needlessly dwelling on it after the necessary investigation is completed.

If you find yourself having adjustment troubles after a shooting, seek out fellow officers who have been through the same thing and discuss your feelings with them. Psychiatrists and psychologists may be of some help, too.

Following my own advice, I'm not going to dwell on this topic. I don't want to blow the problem of adjustment out of proportion. It will generally be as big a problem as you *expect* it to be, and as you *make* of it. It certainly need not be the specter of psychological trauma which many survival theorists who have never been there mistakenly make of it.

FIREARM RETENTION

Approximately 15% of slain officers are shot with their own firearms. That means that nearly 170 officers who were killed over the last 10 years could be alive today, if they had practiced effective firearm retention.

Most of firearm retention technique is in the things we've already discussed, such as disabling and cuffing all arrestees *before* you try to search them, keeping your gun side turned away from people, maintaining your 4' safety interval during interviews, and using a safety holster with a retaining strap. Since an opponent can only get your gun by getting within arm's reach of you, there's also firearm retention value to using your baton and kicks to deliver limited force without getting close enough to be grabbed.

Firearm retention also involves *disengaging* from close encounters. When someone grabs you from behind or from the front, you need to quickly disengage in order to get your firearm out of his arm's reach. Practice the techniques illustrated in the accompanying series of photographs, and any others that you've learned, until you're effective in disengaging from a close opponent to protect your firearm.

In case an opponent succeeds in getting his hand on your pistol grip, don't let him pull on your gun. Grab his hand and hold it in place with your gun hand, pivot your gun side away, and shove the butt of your weak hand hard against the bottom of your opponent's nose with a forceful, upward thrust. Follow through with a kick to his groin or a finger jab to his eyes or throat. Then peel his fingers off the butt of your gun, breaking 1 or 2 of them backwards if necessary to break his grip. Immediately step back out of his reach as soon as you disengage.

In those instances where you draw your sidearm and hold it on a suspect, don't thrust it forward to within his reach. Either step back an extra 2 feet to keep your extended arm away from the suspect, or hold your gun in the protected, close-quarters position. Remember, the suspect can't take away your gun if he can't *reach* it.

FIREARM RETENTION THROUGH DISENGAGING

To stop an advancing adversary, execute this move very quickly.

When grabbed from the front, knee the attacker in the groin.

With arms pinned from behind, stomp hard on the attacker's instep.

With arm free and attacker behind, twist down, cock your elbow, and strike him hard in the face.

FIREARM RETENTION THROUGH DISENGAGING

If the instep stomp fails to break the attacker's grip, simultaneously and forcefully bend at the knees, thrust your arms and shoulders forward, and drive your hips backward into the attacker.

Double up your fist and slam it backward into the attacker's groin.

The threatening attacker grabs you by the collar, while facing you.

Quickly form a "V" with your hands and forcefully bring them up to knock his arm away, while you capture it in the "V" between your hands.

Pivot to the inside and step well across your opponent's front, locking his elbow with your grip on his wrist. →

Kick your feet straight out and then quickly sit straight down in place, while maintaining the pressure lock on the attacker's elbow. The attacker will be forced down with you. If you perform this series of moves rapidly, and forcefully, the attacker's arm will be broken as you sit down. (Use caution in practice.) ↘

One way to respond when your opponent is choking you is to simultaneously clap both hands hard against his ears. This tactic can produce death, so it must be used only when deadly force is justified.

UNDER THE GUN

If a suspect should manage to get the drop on you, your best hope of being able to deflect his firearm and avoid being shot will be to get close to the weapon (if you're not already there), so you can simultaneously pivot away and parry his gun hand aside. Whether he has the gun at your back, your head, or your stomach, it will usually be possible for you to *rapidly* pivot and parry before he can pull the trigger on you (practice with a buddy who's holding an *unloaded* gun on you, and you'll see).

1. Opponent has a gun to your head.

2. Simultaneously pivot in the direction of his gun hand, duck down, and forcefully knock away his gun hand with your fist. Follow through with a knee to his groin as you use an arm lock to direct his weapon toward him.

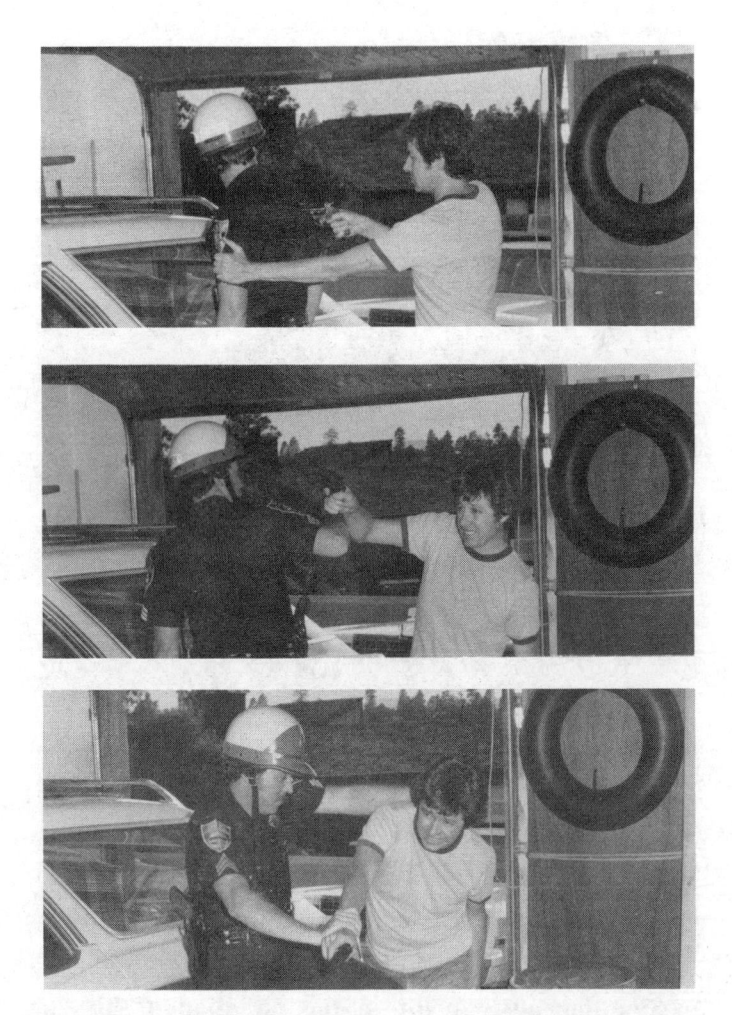

If the attacker has his gun to your back, use the same basic tactics: pivot, duck, knock away the weapon, and follow through to disarm the attacker or redirect his aim so that he is pointing his weapon at himself.

If the criminal is pointing an *uncocked revolver* at your front, within your reach, it may be possible for you to quickly grab the cylinder and hold it tight enough to keep it from turning, which will prevent the revolver from being fired. This maneuver, however, fits into the "desperate trick" category and should not be attempted if you have another alternative (such as pivot and parry). If the crook is strong and fast enough, he can simply pull the revolver straight back out of your grip and blaze away.

This is a desperate trick to be attempted only against an <u>uncocked revolver</u>. Keep your grip firm and kick the crook as hard as you can in the groin. Get out of the line of fire as you try to jerk the revolver from his hand, or turn the barrel back against the outside of his wrist to break his grip.

Whenever you work with a partner, establish a plan of action in the event one of you is being held at gunpoint, while the other is free to shoot. As you know from many cases like the Onion Field incident, you can't submit to armed criminals and depend on their mercy to keep you alive. You need a rehearsed plan to deal with this kind of threat.

The arrangement my partner and I had was that whichever of us was taken hostage would say to the other:

> *Let's do whatever he says. My wife's 8 months pregnant and I don't have a dime's worth of insurance. Don't do anything wrong, please Jesus, this man . . .*

At the word "man," the hostage officer would drop, sag, duck or jump, and the free officer would open fire at the perpetrator.

Actually, neither of us was married. The idea of the pregnant wife was to get sympathy from the gunman. The idea of the lack of insurance was to get the gunman, who would normally have his own financial problems, to identify with the cop's situation. The idea of the plea against doing anything, obviously, was to convince the gunman that we were going to submit. And the idea of saying "Jesus" and then "this man," in almost a prayer-like fashion, was to cause the gunman (who might have a Christian background) to feel that Jesus' attention was being called into focus on the situation—that Jesus was now looking down at "this man."

Have a simple, rehearsed plan worked out with your partner, so that you both can act suddenly, on cue, to neutralize a gunman who manages to take one of you hostage. Use verbal or hand signals, not something which could be misunderstood (like eyeball movements). It's imperative that you and your partner not get your signals crossed.

The key word "man" came in *mid-sentence,* rather than at the end of a completed statement, so the gunman would be waiting for the end of the sentence at the point where we went into action. The hope, of course, was that these little things might put the gunman off guard and induce just enough hesitation on his part to allow us to take him out.

Since we never had occasion to use our plan, I can't tell you whether or not it would work. But at least we *had* a plan—we both knew <u>what</u> we were going to do, and <u>when</u>, if the situation ever arose. That would have given us *some* advantage over a gunman, and *some* is better than *none.*

SAFETY FUNDAMENTALS

The following general, basic information may contribute to your tactical success in particular cases.

Night Vision. More than 64% of officer homicides occur during the hours of darkness. That makes it important for you to be able to see in the dark.

As you may recall from your anatomy classes, the human eye has 2 kinds of cells which are sensitive to light. *Cones* detect colored light; *rods* detect different black-white intensities. To enable you to see in the dark, the rods generate a photopigment called "visual purple." When your supply of visual purple is high, your night vision will be good; when the supply is low, your night vision will be poor.

Visual purple tends to be depleted by lack of rest, inadequate nutrition, emotional stress, exposure to white light, and deficiency of Vitamin A. If you're working a night shift, therefore, it's vital that you preserve your visual purple by getting 7 hours of rest everyday, eating yellow fruits and vegetables and dark green vegetables, practicing stress reduction, avoiding overexposure to white light, and taking a Vitamin A supplement capsule each afternoon or evening. If you go out during sunny days while assigned to a night shift, wear dark sunglasses—several hours of exposure to bright white light can affect your night vision that night.

When you drive at night, don't look into oncoming automobile headlights—look off to the side. If you're suddenly confronted with a glaring white light, shut one eye tightly and keep it closed until the light is gone—this eye will retain night vision and allow you to continue seeing in the dark while the exposed eye begins to regenerate its visual purple.

If possible, keep a second flashlight in your patrol car for inside work, such as report writing, and use a *red* lens on this flashlight. Red light does not destroy visual purple.

In the dark, don't fix your stare at the object you're trying to see—the image will vanish after a few seconds. Instead, use off-center vision and scanning. Let your eyes dart back and forth and up and down to points along the perimeter of the object you're trying to see. This technique will set the image for you, and as long as you keep your eyes scanning in off-center vision, you'll be able to watch the object without focusing directly on it.

Noise Discipline. There may be times when you need to drive up, dismount, and approach a scene without making noise. If so, consider doing the following:
- *Cut your engine and coast into the area.*
- *Be sure your outside PA speaker is off, and turn your inside radio way down.*
- *Use the earplug for your walkie-talkie.*
- *Have your door opened slightly as you coast into place, so you won't have to make noise opening it.*
- *When you stop your car, don't pull the handbrake.*
- *If appropriate, have your sidearm drawn or safety strap unsnapped before getting out of the car.*
- *After you get out, push your car door almost closed, but not enough to engage the latch.*
- *Wear rubber-soled, rubber-heeled shoes that don't squeak.*
- *Never carry more pocket change than a single quarter (for pay phones), so you avoid jingling coins.*
- *Instead of sticking your car keys in your belt, where they can jangle noisily, stick them into your hip pocket.*
- *Hold loose belt items, such as baton and whistle, in place, so they don't slap against each other as you walk.*
- *Keep your leathergear soft and quiet with regular saddle soap treatment to eliminate creaks and squeaks.*
- *Don't chew gum.*
- *Don't talk.*
- *Stay in shape, so running or crawling doesn't start*

you to wheezing like a breathless fat man and so your knees won't pop like a champagne cork when you squat down.

Diversion. Some of the best things to do when you're outnumbered by threatening people, or when you're dealing with a hostile foe, are to create confusion, cause distractions, use surprise, and do the unexpected. All of these things make it harder for your enemy to concentrate on confronting you.

For example, when you're working solo and stop a car full of hostile punks, you may want to make them think you're a 2-man unit. As you pull to a stop, reach over and open your right front door. When you approach the driver, talk as if you have a partner—the one who just got out of that open right front door. Say something like: "The reason we stopped you is because we noticed that ..." If something more is needed, talk to your invisible partner; turn your head slightly toward your car and say: "It looks OK ... just a minute." If the driver you've stopped says he doesn't see your partner, say: "Good. You're not *supposed* to see him."

Obviously, this ploy works better at night than in day. It works better on daytime stops if you can manage the stop at a place where concealment is to the immediate right of your car, so as to suggest that your partner jumped out and got behind the concealment before occupants of the stopped car could see him.

Aside from carstops, you may also be able to use the imaginary partner in other kinds of confrontations by simply acting as if another officer were there. Speak to him, give him hand signals, shake your head, point, or warn him of danger ("Look out, you guys, he's right in there! Wayne, go around the right!'') Some people may be less likely to start trouble with you if they think you've brought reinforcements. And as long as they can't be sure, the confusion will cut down on their options.

Another ploy is to make your suspect think you're interested in something trivial, and not disclose your real interest until you get into a superior position, or reinforcements arrive. For instance, if you make a traffic stop and then recognize the driver as a wanted felon when you walk up, you may be able to reduce his apprehension, and thus his incentive for violence, by acting as if you don't know who he is, or that he's wanted. Make him think he's just going to get a warning on a traffic matter and then be released, and he'll be more likely to play along with you.

You could tell him, for example, that his license plate light is out, or that his rear plate is dangling by a single bolt. Act as if you're a traffic code stickler and this equipment warning is a big deal: "This may seem like a little thing to you, but it's my job to point out defective equipment like this, Sir. I'd like you to step back here and look at it and see if you can get it fixed before you drive on, if you would, please, Sir." When he gets out and heads to the rear of his car, of course, you seek cover and draw down on him.

If you get into a close fight and need a little edge, scream suddenly and loudly as if you'd just gone all-out insane. If you get taken hostage, you might be able to fake a heart attack, or convince your captor you've just snapped under the stress, by talking incoherently about Halloween and your third grade teacher. Anything unexpected and distracting can take your adversary's mind off his objective, can alter his plans, can split his attention, and can give you a chance for survival that you might not get if you submissively play his game, by his rules, doing just what he expects you to do.

OUT OF SIGHT

Once it becomes clear to you and to a suspect that you're interested in him, a basic survival rule is this: *don't let the suspect out of your sight.* If you're making an arrest from a carstop, the suspect may make a move to get something from inside the car, such as his wallet from the glove compartment, or his keys, or his jacket. *Don't allow it.*

There are literally dozens of cases on file where a seemingly-harmless arrestee says to the officer: "Let me get my jacket—it'll only take a second," and comes out with his hand full of a pistol that's already going off in the officer's direction. There are similar cases where a person arrested in his home on a warrant or family fight tells the officer: "Let me get a check for the bail bondsman—it'll only take a second," and comes back with a bucking shotgun.

When a suspect wants to get his hands out of your view for a second, bear in mind that it'll only take a second for him to pull a gun and drop the hammer on you.

I remember being warned about this problem in police academy, and I remember the first time I was thankful I followed the rule that you don't let the suspect out of your sight. I was serving a misdemeanor traffic warrant, and the arrestee who opened the front door turned out to be a striking little redhead, about 20, wearing a short, hard-working, pale green dress. When I showed her the warrant and told her she was going to have to go with me, she said: "But, Officer, I'm not wearing any panties. Do you think I should put some on?"

It went against all my male instincts and contradicted everything I had worked for since junior high school, but I reluctantly agreed with her that she should put her panties on. I started to let her go into the bedroom alone, and then I remembered the rule about not letting suspects out of your sight. When I said: "I'm sorry, Miss, but I have to stay with you," she just smiled back at me over her shoulder.

I followed her into the bedroom and stood guard as she opened her lingerie drawer and poked through an assortment of lacy things, finally selecting a pair of pale green bikini panties, about the same shade as her dress. She said: "I think this color looks good against red hair, don't you?" She turned to face me as she stepped into the panties, pulled the front hem of her dress all the way up to her neck and tucked it under her chin, and took her time working the skimpy underwear up her lovely legs and into place. I had to admit that the color sure looked good against red hair.

From that day on, I always remembered an important rule: never let the suspect out of your sight.

WHAT IF?

During those many quiet hours you spend driving around your district or walking your beat, one constructive thing you can do is to play a mental game of "what if?"

As you cruise past a liquor store or a bank, take a good look at the layout, the surrounding cover, the possible approach avenues, and the door and window configuration. Say to yourself: "What if I answered a call here on a robbery in progress? Which way would be best to come in? Where would I park? Where could I take cover?" Etc. Plan your response, and if you ever *do* get that call, you'll have a head start.

Do the same thing with possible burglary targets and family fight locations. And from time to time, ask yourself: "What would I do if I came under ambush attack right here, right now?" Getting accustomed to figuring out your reactions is an excellent way to prepare for the real thing, and it's an excellent means of keeping yourself constantly survival conscious.

Every time you answer a robbery or burglary alarm that turns out to be false, take a minute when it's over to critique yourself on your performance (unless your supervisor does it for you). Ask yourself whether you used good light and noise discipline on your approach. Did you come by the best route? Did you park in the best place? Did you use cover and concealment properly? Did you take your shotgun with you? Did you observe the situation from a place of cover, or rush in like a suicidal fool? What did you do correctly, and what should you have done differently? What if it hadn't been a false alarm? Would you be dead by now?

DISGUISED WEAPONS

Both in the field and in custody settings, law officers are frequently confronted with an endless variety of disguised weapons, most of which are handmade from common items. Such weapons are designed to be concealed from a casual inspection, and in that regard, they may represent an even more dangerous threat to an officer than clearly-recognizable firearms, knives, clubs, etc.

The following illustrations, presented through the courtesy of the California Department of Justice, are of a number of commonly-encountered disguised weapons, to illustrate how ingenious your enemy can be at devising weapons to use against you. This is not an exhaustive encyclopedia of such weapons, but should be a sufficient sample to show you that just about *anything* in a criminal's possession could be a disguised weapon, and deserves your closest scrutiny.

CONTROLLED ACCESS INFORMATION. FOR LAW ENFORCEMENT USE ONLY. DO NOT COPY.

SURVIVING FIELD THREATS/303

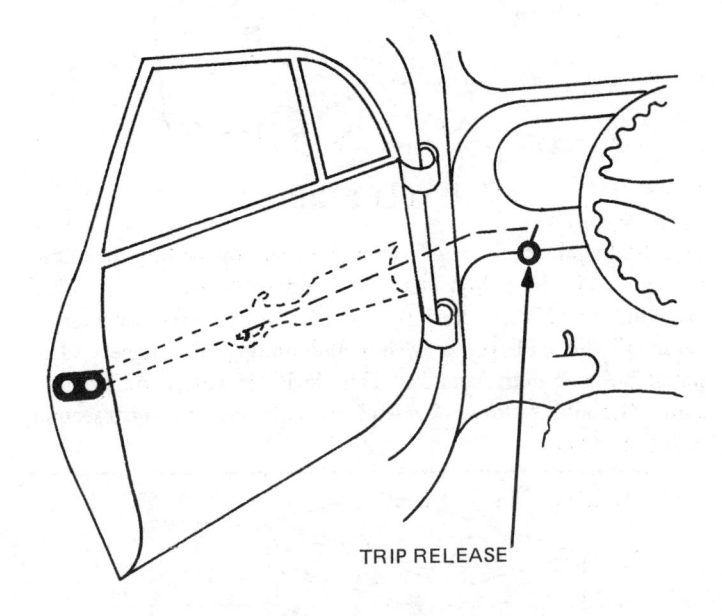

TRIP RELEASE

CAR DOOR SHOTGUN

A modified 12-gauge shotgun is mounted inside the driver's door of the criminal's car, with muzzle pointed rearward. As the officer walks toward the driver's door from the rear, the cop-killer opens the door slightly, pulls a trip wire concealed in the dash, and cuts the unsuspecting officer in two. This disguised weapon is specifically designed for murdering law officers.

Your best protection against this menace, most often reported around the New York area, is to watch for opening car doors and evade, and to keep your approaches as close to the suspect vehicle as possible.

WALLET GUN

This folding leather wallet, with a trigger-finger hole near the center, conceals a high-standard Model DM-101, .22 caliber magnum, or .22-LR Derringer, which can be fired by merely "aiming" the end of the wallet and pulling the trigger. The wallet holds 2 extra rounds. This device is widely available by mail order and in stores. Remember, "It'll only take me a second to get my wallet."

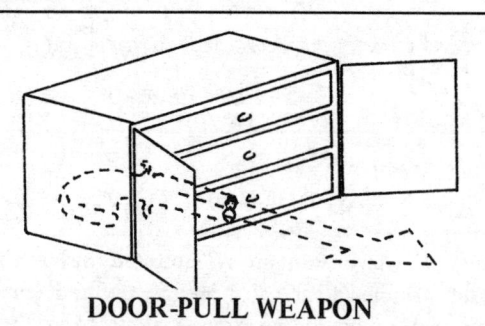

DOOR-PULL WEAPON

Criminals with reason to expect searches of their homes, such as narcotics dealers, often rig refrigerator doors, dresser drawers, desk drawers, bureau drawers, etc., with pre-aimed weapons connected by wire or string to the drawer pull or door. When the searching officer pulls the door or drawer open, he shoots himself (illustrated here: a derringer in a jewelry box). To protect yourself, stand well to the side when opening doors or drawers in a suspect's residence, or use a long line to keep out of range of firearms and explosives.

SURVIVING FIELD THREATS/305

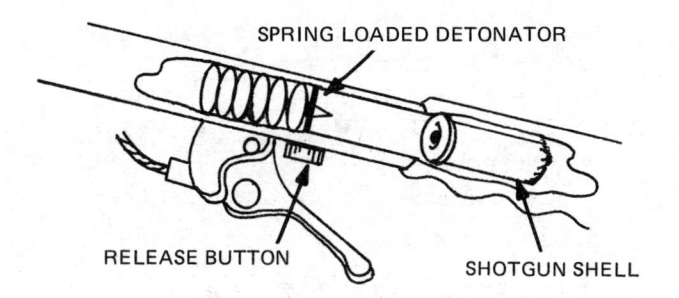

HANDLEBAR SHOTGUN

The biker's concealed shotgun, which can be aimed by turning handlebar and leaning the bike, is fired by releasing a spring-loaded firing pin, driving it against the base of the shell. Watch for open-end handlegrips, and unusual maneuvering of the handlebars as you approach a stopped biker.

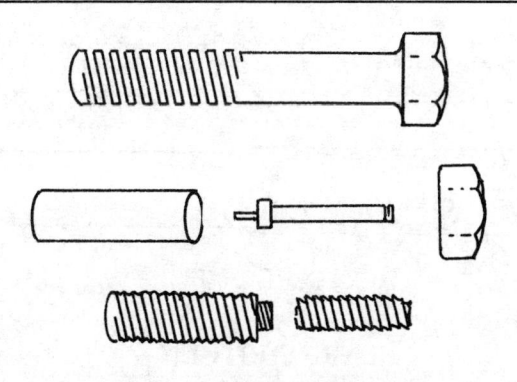

BOLT GUN

Another biker specialty, this 5/8-inch machine bolt is fitted into the Harley-Davidson frame, while loaded with a .22 long rifle hollow point cartridge. Once removed, it is held in the hand and fired by pulling back and releasing the spring-loaded hex bolt head. Distrust anything held in a suspect's hand.

306/THE OFFICER SURVIVAL MANUAL

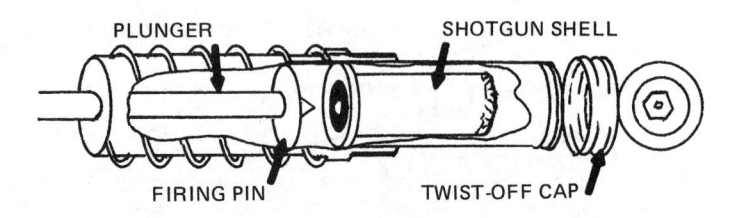

SHOCK ABSORBER SHOTGUN

A motorcycle shock absorber is loaded with a shotgun shell and fired by manual pump action of the shock.

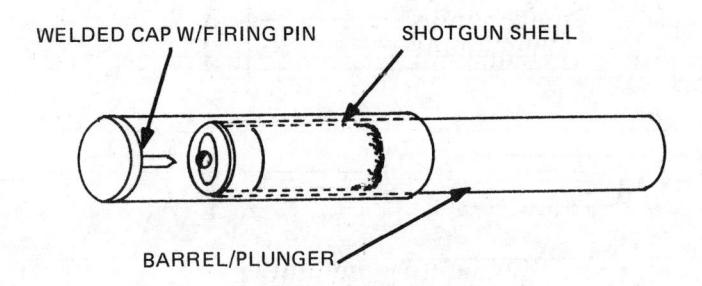

FLARE SHOTGUN

A device that resembles a red highway flare at first glance, actually consists of a barrel loaded with a shotgun shell, inside a detonator sleeve. The criminal holds the closed end firmly against his car or motorcycle while aiming the "flare" at the target, and then plunges the loaded barrel back against the firing pin. "Let me get a flare—it'll only take a second."

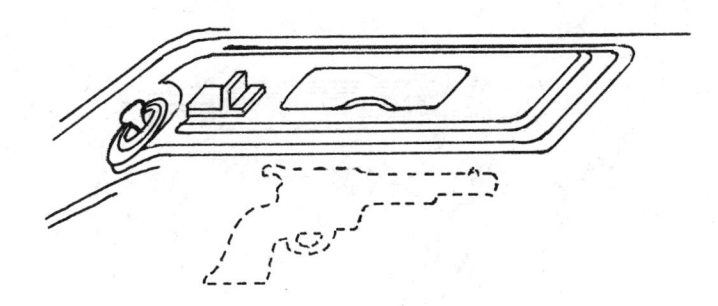

LOADED ARMREST

Entire plate on top of the vehicle armrest pops up and out when the ashtray lid is pulled up, giving the driver or passenger immediate access to a concealed handgun. Be alert to occupants of a car whose hands rest on and tinker with the armrest cover.

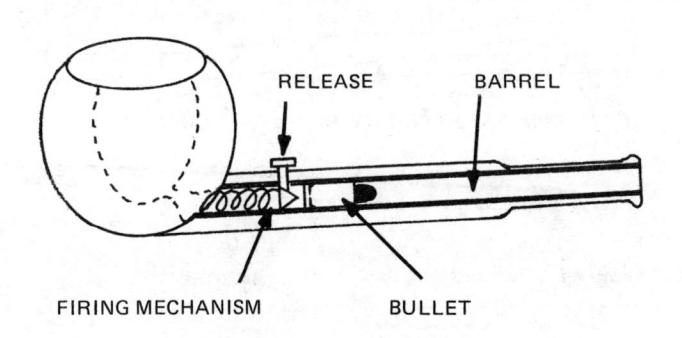

SMOKING PIPE GUN

A smoking pipe with a bored stem, a spring-loaded firing pin, and a .22 or .32 caliber round. A thumb-release button on the stem, near the bowl, is the only visible clue to this deadly mechanism, and is easily concealed beneath the criminal's thumb.

308/THE OFFICER SURVIVAL MANUAL

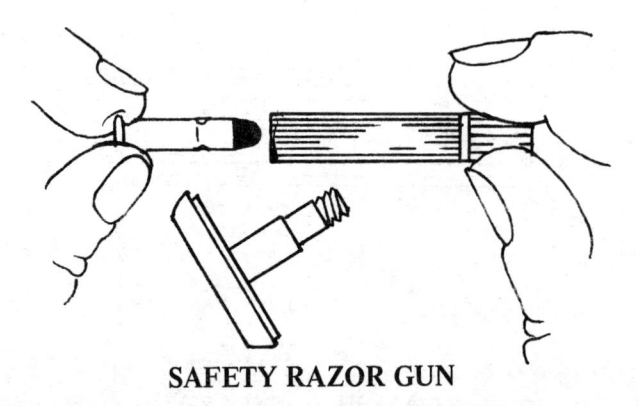

SAFETY RAZOR GUN

Often found in jails and prisons, this razor gun features a hollow handle which serves as a barrel for a .38 round, detonated by striking the primer against a crude firing pin made into the blade end portion of the razor (assembled razor is struck hard against a solid surface).

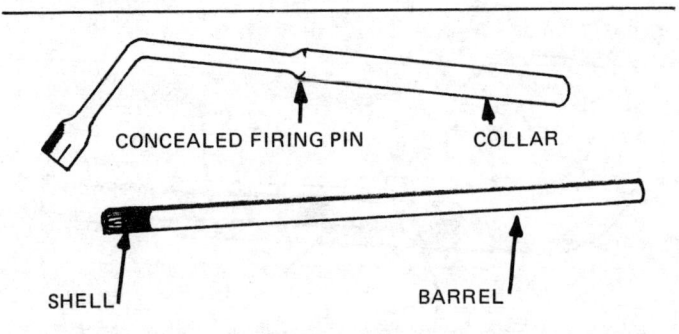

TIRE IRON SHOTGUN

A single-shot, 410-gauge shotgun, concealed in a 2-part tire iron, most frequently reported around Miami, Florida. A shell loaded into the cylindrical "barrel" is fired out the open end when the barrel is inserted into the tire iron collar and jammed back forcefully against an imbedded firing pin.

SURVIVING FIELD THREATS/309

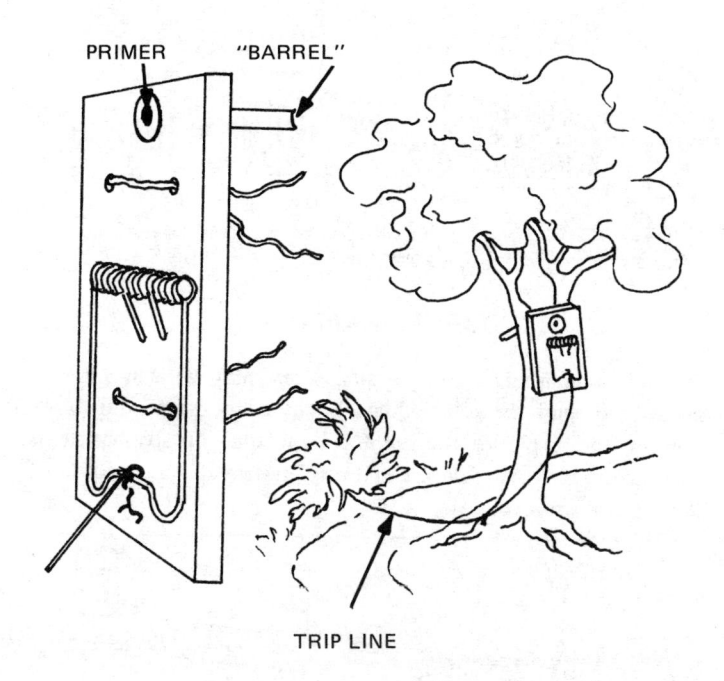

Diagram labels: PRIMER, "BARREL", TRIP LINE

RAT TRAP BOOBY TRAP

Used around marijuana patches, this device is activated by a trip wire across trails or other points of entry. The rat trap is loaded with a .22 caliber magnum round, seated in a small section of ¼-inch copper tubing, which serves as the barrel. Trip release of the spring action drives a firing point against the primer, firing the pre-aimed round at the person crossing the trip wire.

Indoors and out, always watch for trip devices across your path. Avoid activating these devices, secure the area, and call for the bomb squad. If you're forced to deal with such devices yourself, tie a long string to them and get behind cover before activating.

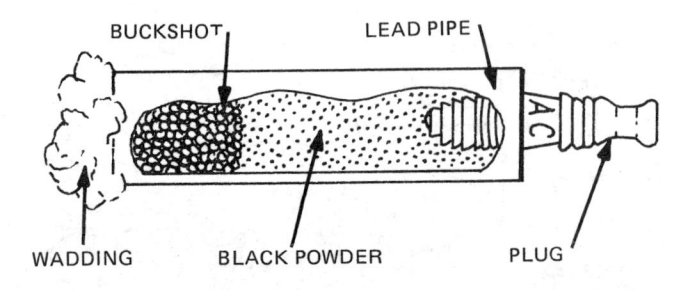

VEHICLE EXPLOSIVE

When this bomb is attached to a spark plug wire in a car or motorcycle and the vehicle started, the spark ignites the black powder and explodes the bomb. Depending on size, contents and placement, this device can easily destroy the occupants of the car or motorcycle.

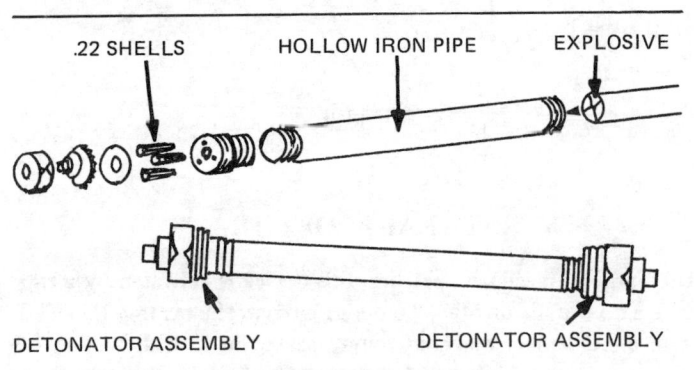

TORPEDO PIPE BOMB

This device has a high explosive filler in the center, and detonators on both ends. When thrown from a high building, or hurled in an upward arc to land on the highway in front of a pursuing patrol vehicle, the torpedo lands on either end, firing .22 shells into the explosive, and exploding the iron pipe into deadly fragments.

STREAMERS BALL BEARING ATTACHED W/EPOXY

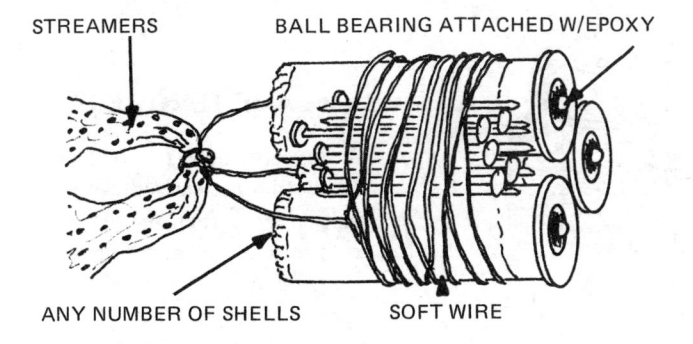

ANY NUMBER OF SHELLS SOFT WIRE

SHOTGUN SHELL BOMB

A small "parachute" attached to the top of this throwing device assures that when thrown from a multi-story building or tossed from a pursued vehicle, the ball bearings will strike the street, detonating the bomb, and sending buckshot, pieces of nails, and wire fragments in all directions.

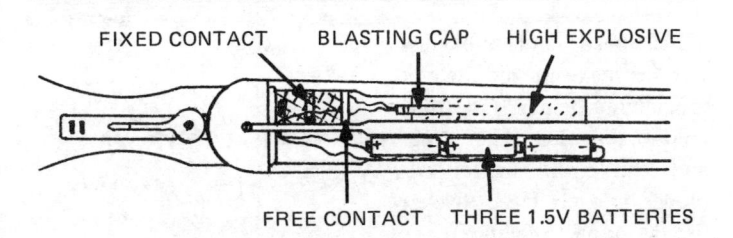

FIXED CONTACT BLASTING CAP HIGH EXPLOSIVE

FREE CONTACT THREE 1.5V BATTERIES

EXPLODING SHOTGUN

Left in an obvious place to be confiscated by officers, a device like this exploded and killed 2 officers who were tampering with it at a police station. Breaking the weapon open brings the free contact into touch with the fixed contact, closing a circuit between the batteries and blasting cap.

FLAME THROWER

A cigarette lighter is tied or taped to a pressurized can of hair spray. When the lighter is lit with a high flame, spraying a lacquer-base hair spray through the flame throws an intense flame up to 12 feet.

CIGARETTE LIGHTER EXPLOSIVE

Match heads and smokeless powder make up the filler in this altered lighter. The flint ignites the firecracker fuse, and the entire lighter explodes in your face. (Never use the suspect's lighter.)

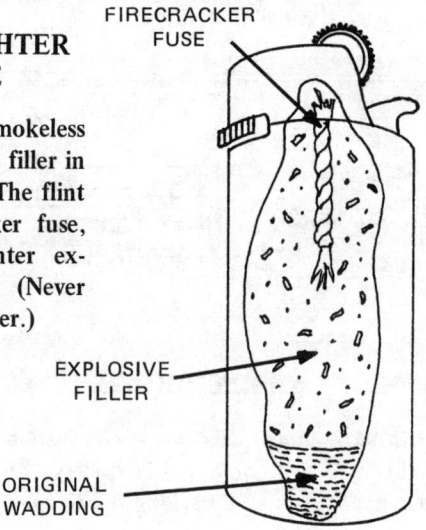

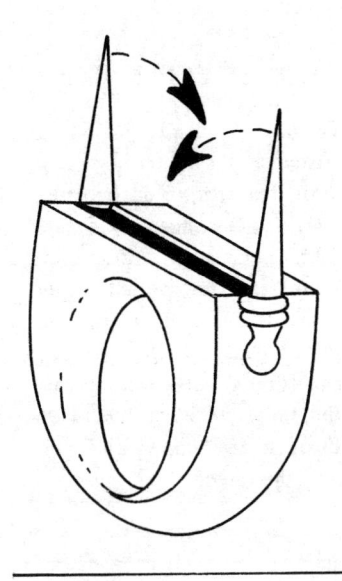

"GUARDIAN RING"

The sharp metal projections on this ring are spring-loaded to pop out with slight thumb pressure. When not in use, the miniature swords fold down flat into the face of the ring.

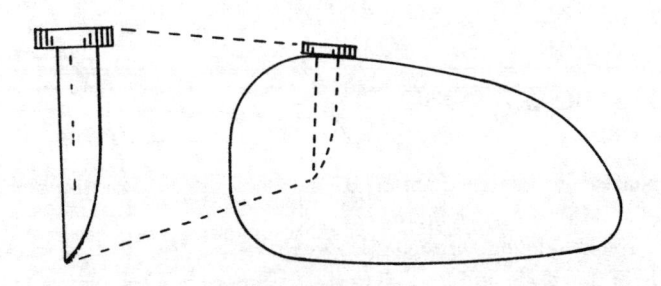

GAS CAP DAGGER

A knife blade is brazed or welded to the bottom of a motorcycle gas cap. This dagger is normally concealed, but can be quickly removed by the rider and used against you. Watch the rider's hands.

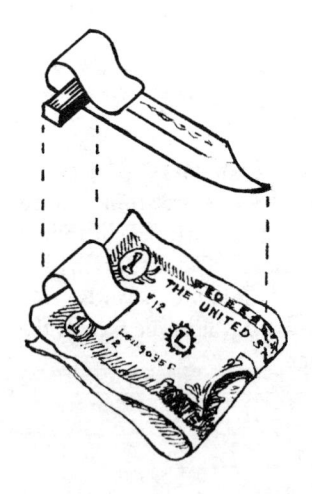

MONEY CLIP KNIFE

Turning up from Florida to Wisconsin, the money clip knife features a 3-inch blade, which is concealed inside folded currency. The clip-knife can be pulled away from the currency, or the folded end of the currency can simply be shoved against the target, driving the blade through the paper bills and into the target.

DASHBOARD HOOK

A concealed weapon common in the Midwest resembles a hay hook, hooked through a hole in the dashboard, so that only the handle is visible. It can quickly be removed by pulling out and down, and the steel hook becomes a dangerous weapon.

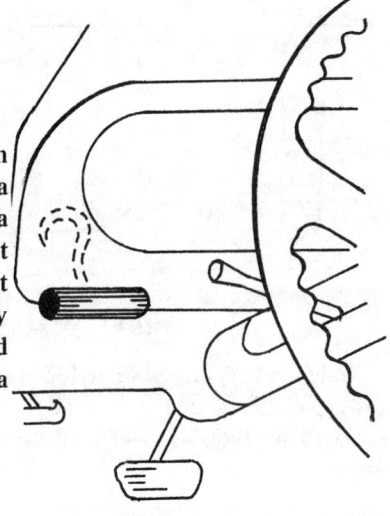

BELT BUCKLE KNIVES

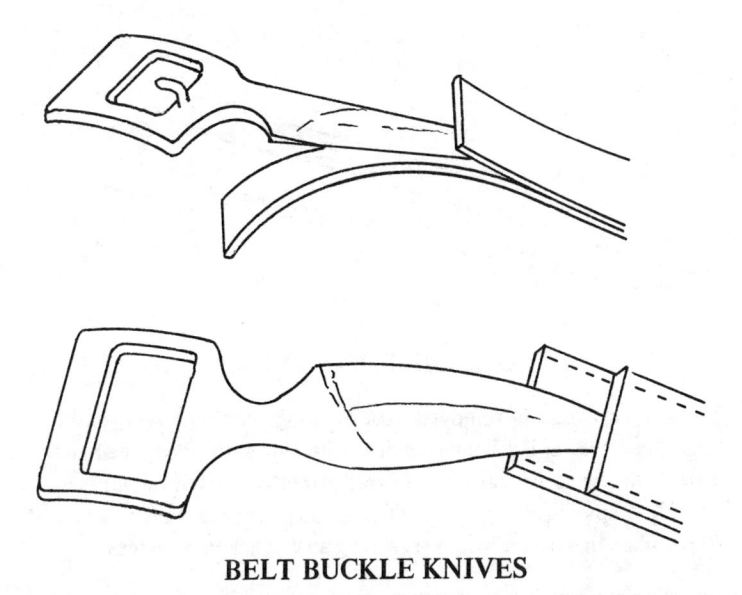

A wide variety of belt buckle knives are sold in sporting goods stores and gun shops. These knives may be single edged (top) or double edged (bottom), and usually are 3" or shorter. They are removed simply by pulling them out of the belt-sheath.

COMB STABBER

Advertised and sold as "The Defender," this plastic comb includes a spring loaded thin steel shaft, similar to an ice-pick. When the button on the end (right) is depressed, the shaft springs out the other end. This item comes in either a black or red case, and sells for about $5.

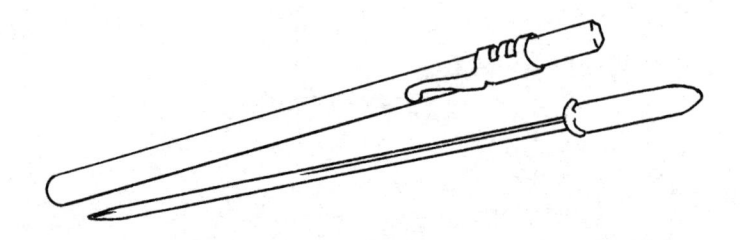

BALLPOINT PEN KNIFE

The ink cartridge is removed from a long ballpoint pen, and a large needle or spike is affixed into the cap with epoxy cement. This device is easily carried without attracting attention, and can be pulled apart quickly to produce a dangerous stabbing weapon. Remember this device when searching suspects and arrestees.

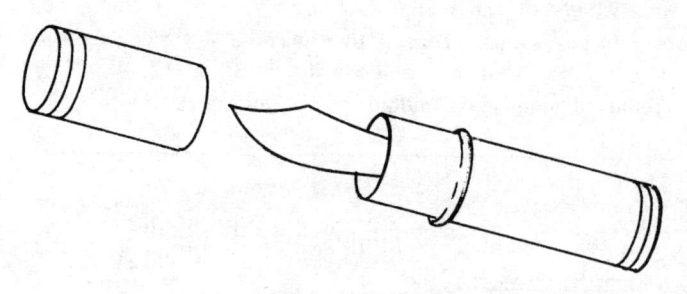

LIPSTICK KNIFE

The 1½-inch blade contained in this lipstick container (taken from a woman arrested in Louisiana) is extended by turning the base, just as you would normally do to extend a lipstick. When the blade is retracted and the cap closed, this item looks just as harmless as all those lipsticks you may have bypassed in your previous searches of women.

DINNER FORK "KNUCKLES"

A frequent threat in jails and prisons, the dinner fork is simply bent into the position illustrated. Worn on the outside of the fingers, as shown, it is an obvious stabbing device. It can also be worn with the prongs on the inside of the cupped hand, and used to injure someone in an otherwise-harmless slap.

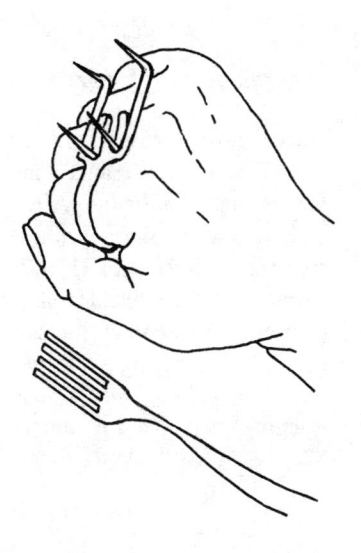

KNUCKLE KNIFE

This weapon consists of a 1/8-inch thick blade of stainless steel, affixed to a ring which is worn over the middle finger. This knife is easily mingled with a suspect's pocket change. It is sold by mail-order for about $50.

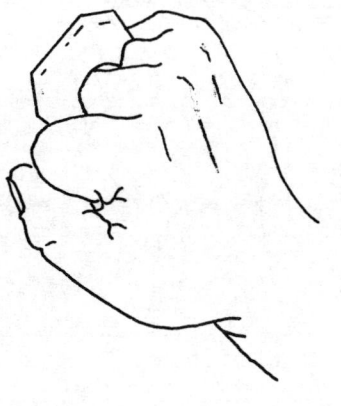

METAL HAIR PICK

Plastic devices similar to this one in shape are often used in hairstyling, especially Afro styles. Razor sharp prongs cut from sheet metal, and worn over the knuckles, are painted to resemble plastic hair picks, and may escape detection in long hair. These weapons are sold for about $25.

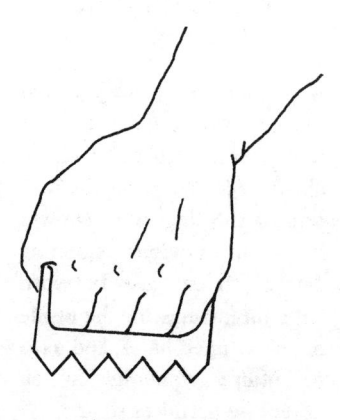

KNUCKLES OF NAILS

This device consists of 2 layers of duct tape covering a row of 6d. or 8d. nails, the sharp points of which protrude about ¼ inch. Wrapped around the knuckles as shown, this item can be used to draw a lot of blood with a single blow. Since it is flexible, it can be rolled up, folded from end to end, or stretched out straight and concealed inside clothing.

FOLDING HUNTER

The "Atchisson Folding Hunter" is a 4-blade folding knife made of carbon steel. It folds into a compact 7 inches, or unfolds into 12 inches by 12 inches. Held by one of the blades and thrown, it will penetrate 3/4-inch plywood at 20 yards. This weapon sells for $14.95.

CHINESE SPINWHEEL

This handmade version is formed by welding the heads of eight 16d. nails together in a circle. Smaller nails can be used, of course, for spinwheels of smaller diameters. If thrown forcefully in a straight line, these devices can be deadly accurate.

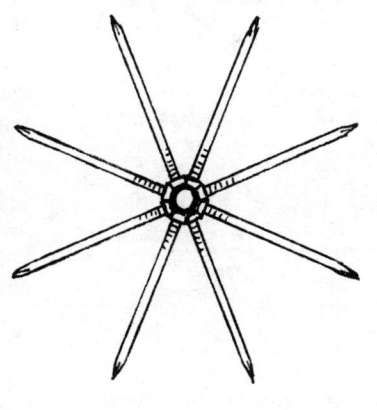

JAPANESE THROWING KNIVES

Two varieties of the "Shuriken," made from steel or sheet metal, and readily available in many sporting goods stores. These items are sometimes carried in pockets, or incorporated into belt buckles, or worn as neck medallions on chains.

The hole in the center of the Shuriken gives it aerodynamic stability and high velocity, enabling the thrower to imbed the spinning blades deep into a target.

Although these items may look ornamental when worn by a person, they are potentially dangerous weapons which you should treat as any other dangerous weapon.

YOUR OWN WORST ENEMY

During the last decade, the FBI has been able to draw a statistical profile of the people who kill cops. Out of 100 cop-killers, 8 will be juveniles; 64 will be aged 18 to 30; and the remaining 28 will be over 30 years old.

Only 4 will be female, and 96 will be male. Half will be black, and the remaining half will be white, Latin, and other races; 72 of the 100 will have prior criminal records.

Of every 100 offenders arrested and charged for killing officers, only 63 will be convicted of murder. Of every 100 offenders convicted in an officer murder, only 8 will receive the death sentence, and based on statistics for the last 10 years, not one cop-killer will be executed in the United States.

You've got to watch out for your own survival, because not only will the criminal justice system not bring you back—it won't even do justice to your murderer! And not withstanding the FBI's statistical profile of your killer, you and I both know that your worst enemy is the cop inside you—the one who wants to be macho, who wants to be cool, who falls into dangerous routines, who takes your survival for granted, who lapses into complacency, and who thinks that nothing bad could ever conceivably happen to you.

That macho-complacent cop is the biggest threat to your survival. He's the one you have to fight the hardest, and the most often. He's the one you have to whip if you're going to meet your survival goal. If you end up in a hole in the ground next Monday—no matter which way it happens—that cop inside who kept you from practicing good survival habits is going to be more to blame than anyone else. If you don't take every aspect of your survival seriously, *you* are your own worst enemy.

On the other hand, if you convince that careless cop inside to make survival your full-time, number one objective, your chances of survival are excellent. If you'll take the time and the effort to follow sound tactics, you can be your own best defense.

I told you earlier that I had examined a number of the leading officer survival publications, and as you've noticed, I've frequently disagreed with some of them. My *strongest* disagreement is with the assertion in one of those books that if somebody wants to kill you, you're helpless to prevent it, no matter how good your training, tactics, physical condition and protective equipment may be.

That conclusion is absurd, of course, because if you couldn't possibly avoid being killed by someone intent on murdering you, there would be no reason for anyone—including that particular author—to write a book advising you how to do it. There would be no point in firearms training, or tactical instruction, or any other training designed to help you survive hostilities.

You *can* prevent your own murder. You *do* have control over your own survival. You do not have to accept the defeatist notion that if someone is determined to kill you, he will. There are hundreds of officers walking around alive today who were attacked by murderous assailants, but who survived and even destroyed their attackers. These officers are living proof that you don't have to give up and die, just because someone wants to kill you. *You can survive* field threats! ◻

NOTES ON LOCAL RULES

8

SURVIVING WOUNDS AND INJURY

If you *do* find yourself with life-threatening wounds or injuries as the result of a traffic accident, a shooting, or an assault, you need to know how to treat yourself (or how to direct others to help you if you're incapacitated) until medical aid is available. *If you don't know what to do to save your own life, you may die needlessly of respiratory failure, shock, or loss of blood, even though your wound or injury was not serious enough to cause your death.* Meeting your survival goal requires that you learn how to survive the effects of trauma with emergency self-aid. (These basics may also come in handy when you're the only hope of saving your fellow officer's life, or the life of a member of your household.)

326/THE OFFICER SURVIVAL MANUAL

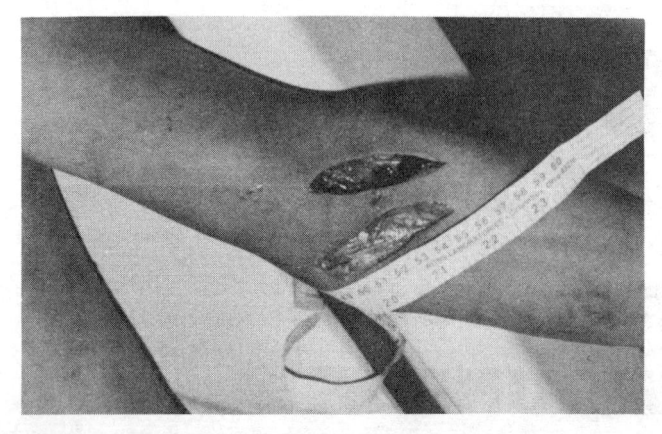

Self-aid measures could have saved this officer's life. Instead, he needlessly died of shock and uncontrolled bleeding after sustaining two deep lacerations.

The following techniques are simple, *first-aid* measures to be taken when imminent loss of life or limb demands *immediate* action, and trained medical personnel are not on the scene. (In case of any serious injury, it is extremely important to get to a hospital as soon as possible.) If you're trapped in a wrecked patrol car in the middle of nowhere at 4 in the morning with your left foot missing, or you're pinned down by gunfire while your blood runs across the street and into the storm drain, or you're lying in the back seat of your patrol car with a sucking chest wound while your partner rushes you toward a hospital that may be only 2 blocks beyond the place where you'll take your last gasp of breath, you can't pull this book out of your briefcase and find out what to do to keep from being next Monday's funeral. **Now,** while you have that chance, learn how to prevent a *serious* injury from becoming a *fatal* one. All it takes are 4 lifesaving steps.

STEP 1: DON'T PANIC

You can't carry out steps 2, 3 and 4 if you lose your head. Don't go into a panic just because you feel a lot of pain or see a lot of blood. If you break your femur (thigh bone), you'll have terrible pain, but you can survive this injury very easily. If you take a grazing shot to the scalp, you'll see a lot of blood, but your life may not be seriously threatened, provided you take prompt first-aid steps. Just because you get hurt, don't immediately say to yourself: "Oh, God! I'm going to die!" Instead, say to yourself: "Thank God I'm still alive! I'm going to make it!"

Remember that the odds are in your favor if you're still alive after the injury. For every person who dies of traffic accident injuries, there are 57 other injured people who survive. For every officer who dies of wounds inflicted by assault, there are 203 other injured officers who survive. Those are good odds. Keep them in mind when you get hurt.

If you feel like you need calming and there's no one around to talk to, talk out loud to yourself—even the sound of your own voice can reassure you. Say to yourself: "Maybe that crazy Rutledge was right. OK, I've survived this far and I've got Step 1 under control—no panic. Now, what was Step 2?"

STEP 2: STOP THE BLEEDING

Depending on your weight, you probably have approximately 6 to 7 quarts of blood. Loss of more than 1 quart of blood is a threat to your survival. Don't let it run out onto the street. Take the following lifesaving measures, in *this* order:

Apply Direct Pressure. Using a compress from your first-aid kit, or a clean handkerchief or piece of your T-shirt, or, if you don't have anything else, the palm of your hand, apply firm pressure directly over the wound. The purpose of this is to stop the blood flow long enough to permit clotting.

If your compress becomes blood-soaked, *don't* take it off and throw it away—add more material or apply your hand directly on top of the bloody compress. If you pull it away, you may loosen clots that were just beginning to form. Keep pressure on until you get help.

(Even if you've never had a runny nose in your life, I recommend that you never go to work without a clean handkerchief—not Kleenex—in your back pocket. You may not be able to get into the trunk and pull out your first-aid kit when you need that compress.)

Elevate The Part. *If there is no fracture,* raise the injured arm or leg above heart level so that gravity assists in reducing the blood pressure at the site of the wound, thereby slowing the loss of blood. If your injury is to the shoulders, neck, or head *(and no*

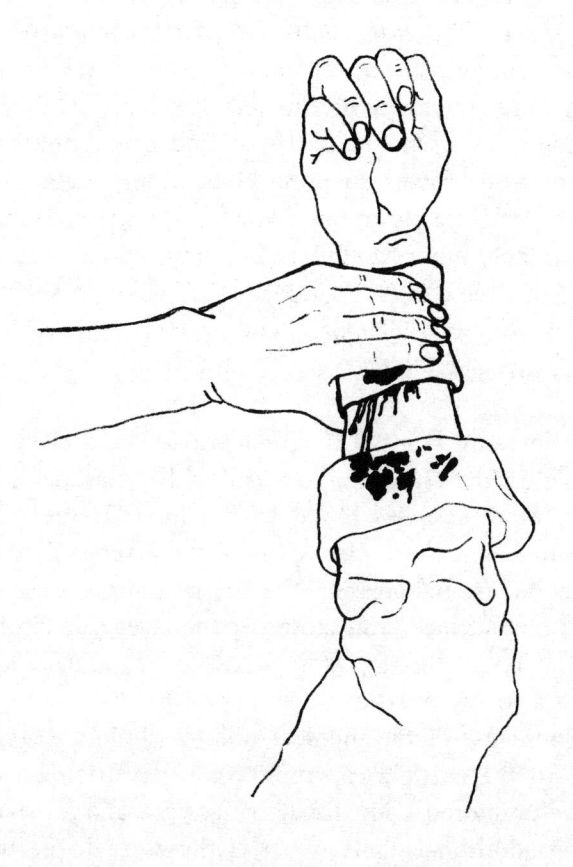

To stop the bleeding, apply direct pressure right on the wound with a compress or your hand. If possible, keep the wounded area elevated to slow the flow of blood. Do not remove a blood-soaked compress.

fracture), try to get into a sitting position to elevate these areas. *Do not relieve the direct pressure*—use pressure and elevation together to stop the bleeding.

If you're wounded in the left hand and still busy shooting with your right, try to find something solid (part of your cover) to press your wound against for pressure, and use elevation if you can. You may simply have to hold your wound tight against your chest or along your leg. It won't do you any good to keep up the gunfire if you let more than a quart of your blood run out through a hole.

Use Pressure Points. If direct pressure and elevation fail to stop the bleeding, you can apply pressure on the artery which supplies blood to the injured area. The "pressure points" are places where the arteries lie close to the bones; by pressing the artery tight against the hard, bony surface, you close off the artery, much like stepping on a running garden hose on a hard surface, such as a concrete patio.

Because use of the pressure point technique cuts off blood to the entire area supplied by that artery, it can deprive uninjured body tissues of oxygen and nutrients, causing additional damage. It is therefore to be used *only when* direct pressure on the wound and elevation fail to stop the bleeding. The pressure point is to be used *only as long as necessary* to stop the bleeding. It is to be used only for wounds of the hands, arms, feet and legs—not for neck, head or trunk wounds.

There are 4 pressure points on your body for controlling the blood supply to your extremities. The pressure point for each arm is on the inside of your upper arm, between your biceps and triceps muscles, about midway between your elbow and armpit. To cut off the blood flow to your arm or hand wound, rest your injured arm in the V between the thumb and fingers of your good hand (fingers on the inside of your arm, thumb on the outside). With the flat insides of your fingers (not just the fingertips), press hard against the bone in your arm.

After the first couple of minutes, ease the pressure slightly to see whether the bleeding resumes. If it doesn't, relieve the pressure point and continue with elevation and direct pressure. If the bleeding resumes, continue using the pressure point, checking again every minute or so until the bleeding stops or you get medical help.

The pressure point for each leg is at the front of the place where your thigh bone connects to your pelvis. This is the place where you will see a crease if you lie on your back and raise your leg up into the air. It is generally about the place where the legband of your jockey shorts crosses the front of your thigh. To cut off the blood flow to your leg or foot wound, double up your fist (same side of your body as the wound, if possible) and press it hard against the pressure point. If your other hand is free, place it on top of your fist to apply additional pressure.

If you are directing someone else who is giving you first aid, lie on your back and have the person place the heel of his hand against your leg pressure point, with his arm locked, leaning toward you. Remember

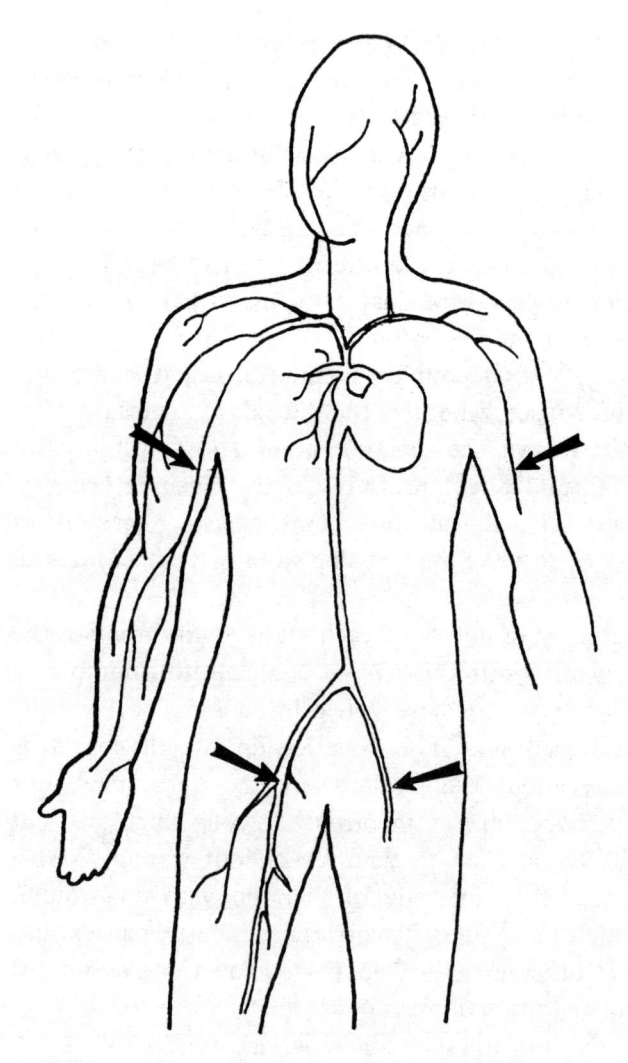

Pressure points are indicated by the arrows. Press firmly at these points to cut off blood supply to an injured limb.

to keep the wound elevated and maintain direct pressure. Don't use the pressure point any longer than necessary to stop the bleeding.

Apply a Tourniquet. If direct pressure, elevation and pressure point technique do not stop the bleeding (they almost always will), and if you face a choice between losing your life and losing an arm or a leg (tourniquets are that dangerous), apply a tourniquet as a last-resort, lifesaving measure. You should not need to resort to tourniquets if you are bleeding from a small puncture wound (bullet or ice pick hole), a shallow laceration (sliced with knife or glass), or the traumatic amputation of a digit (finger or toe shot off, or sheared off in a traffic accident).

If the other techniques don't control the bleeding, you may need to use a tourniquet if you suffer a massive gunshot wound (large-caliber, high-velocity round, or shotgun blast), a deep, jagged incision (from knife attack or traffic accident), the traumatic amputation of a hand, arm, foot or leg (from gunshot, accident or explosion), or a large avulsion (chunk of your body torn away, as by gunshot, accident or explosion), or if a major limb artery is severed. Tourniquets are only for injuries to the extremities—not for wounds to the trunk, neck or head.

To apply a tourniquet, use a flat band of cloth at least 1" wide (do not use narrow strips, ropes, strings, or wires—these can cut through your tissue). If nothing better is available, use your necktie, rifle sling, handbag strap, or belt, or tear a band of cloth from your uniform shirt (T-shirts and pantyhose tend to roll up—you want something that stays flat). Wrap the band around your arm or leg *twice, just above the wound.* Tie 2 overhand knots and lay a strong stick across the knots (use your baton or flashlight if necessary). Tie 2 more overhand knots (you may have to use your teeth and one good hand to tie the knots).

Tighten the tourniquet by twisting the stick around in a circle until the bleeding stops. *Tie the stick in place* along your arm or leg with another band of cloth (not too tight—just snug enough to secure the stick).

Once it is in place, **do not loosen or remove the tourniquet**—this will probably start the blood loss again and create a high risk of the immediate onset of shock. The tourniquet is to be removed only by a doctor.

Leave your sleeve or pant leg rolled up above the tourniquet and do not cover the tourniquet with anything. Using your ballpoint pen, write the time you applied the tourniquet right on your skin, just above the tourniquet. Then write a large "T" on your forehead, and if you can, write a note with the location of the tourniquet and the time applied, and pin this note to your shirt or jacket with your badge (if you have no pen, write the "T" on your forehead with your finger, using some of your lost blood). If you lose consciousness before you reach the hospital, medical personnel will see the "T" or the note and begin searching for your tourniquet. The sooner the doctor can

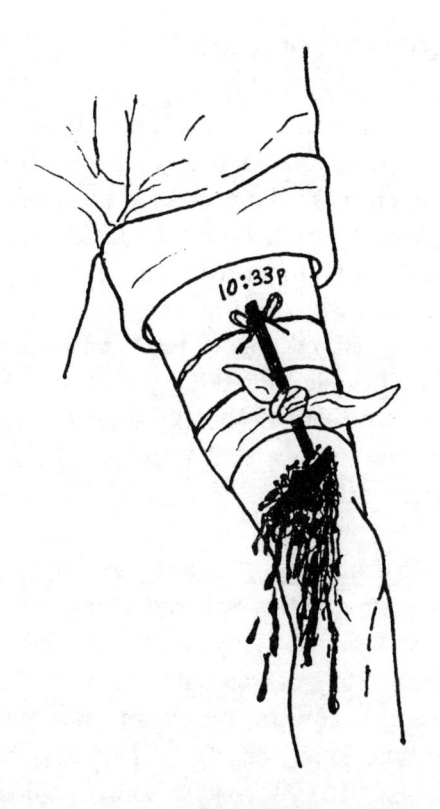

THE TOURNIQUET

As a last-resort measure to stop the bleeding, put the tourniquet just above the wound. Always try to save knee or elbow, if possible, to make it easier to use a prosthetic limb, if you lose the lower arm or leg. Once it's in place, do not loosen or remove the tourniquet.

remove the tourniquet, the less damage your lower limb will suffer, and the less likely you will lose the limb.

STEP 3: PROTECT YOUR AIR

Your brain will suffer irreversible damage if it is deprived of oxygen for 4 minutes. You will probably die if your brain lacks oxygen for 6 minutes or more. If you've suffered chest, throat or facial injuries that threaten your ability to breathe, and if there's no one around to care for you, you must take steps to protect your air supply in case you lose consciousness. If you're having trouble breathing, or if you see your skin starting to turn blue (check beneath your fingernails), take steps to clear and keep your airway open, and your lungs working.

Remove Obstructions. If you wear dentures, get them out of your mouth. Spit out your chewing gum, loose teeth, and blood. If your tongue isn't free and forward, reach into your throat with your finger and pull your tongue toward the front of your mouth. Get your necktie loose, or off. Unbutton your collar and any tight buttons on your shirt. Loosen your belt and waistband. Loosen body armor or bra straps, if necessary. Unsnap your helmet chin strap.

Position Yourself. The nature and location of your injuries will dictate whether you sit up or lie down. In either case, you need to hyperextend your neck by tilting your head back slightly and keeping your chin

With your neck straight, your airway may become sealed off if you lose consciousness.

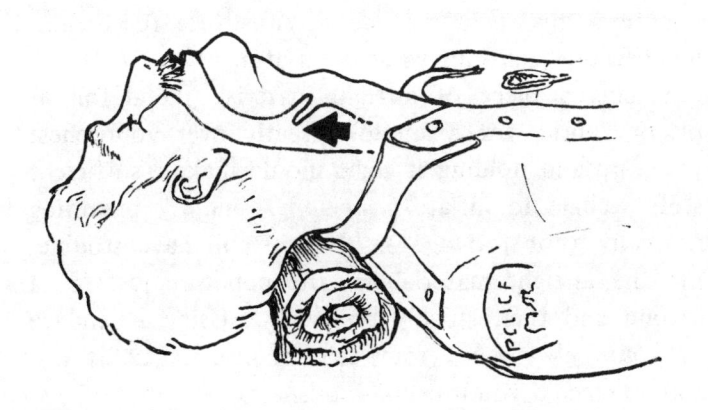

Hyperextend your neck to keep your airway open.

up. This will keep your airway open, even if you lose consciousness. If sitting, try to lean your head back over some stationary object so it will be supported if you pass out (don't want to slump forward). If lying on your back, place an object (rolled blanket, jacket, softhat) beneath your neck—not under your head. If blood or vomit is draining from your mouth, turn your head to the injured side so that the fluid can drain from the corner of your mouth.

Seal Your Lungs. The air pressure inside your lungs is lower than the outside air pressure. If one of your lungs is punctured, the higher outside pressure will cause that lung to collapse, and the other lung may also become compressed. To restore the pressure within your lungs and get them working again, you need to seal the puncture (this is called a "sucking chest wound," because your lung sucks air through the puncture—this sucking is not a substitute for breathing, because your lung will not inflate).

Placing a piece of airtight material (metal foil or plastic fabric, for example) directly over your chest puncture and holding it tight should make a sufficient seal. Inhale to inflate your lungs, and try breathing normally (not too deep). If you still have trouble, lift the airtight material off the puncture just for a second and then quickly replace it. Using a bandage or your belt, secure your airtight material so it will stay in place if you lose consciousness.

(I once saw a fellow soldier stick his finger into a sucking bullet hole in his chest in the middle of a heavy gunfight and keep on shooting. If you have to, you can do that, too.)

STEP 4: PREVENT SHOCK

Shock is a state of severe depression of your body's vital processes. It's possible to die of shock, even from relatively minor, non-fatal injuries. Shock is a serious threat to your survival following a wound or injury. Even if you don't think your injury is serious and you feel calm and collected, take steps to prevent shock—once it develops, it is difficult to reverse.

Shock may be caused by a rapid loss of blood or oxygen, by heart attack or stroke, by perforation of a stomach ulcer, or even by the loss of large amounts of body fluid, such as through vomiting or wound drainage. To prevent shock, you need to take the steps we've already discussed for improving circulation and breathing, as well as the 2 following measures.

Maintain Body Temperatures. When your vital activities are depressed, your body temperature starts downward, giving you chills, and making the shock worse. To prevent the loss of body heat, try to lie on something other than cold ground or pavement. If you're able to move, try to get to your car seat, or find a board or a piece of cardboard or discarded newspaper to put between you and the ground. If you have a blanket (or are directing another to aid you who can get one), put part of it *beneath* you and part of it *over* you.

Do not pile on excessive covers. This will overheat your skin and divert your blood supply from vital organs to the skin. All you want to do is maintain your normal body temperature.

Replace Body Fluids. If you've lost large amounts of fluids and you have access to water, drink 1 cup every 10 or 15 minutes, *unless you have stomach or abdominal injuries.* In spite of what you've seen on TV, **never** drink alcoholic beverages when you're wounded or injured. Alcohol depresses your systems and can trigger shock.

If you have no chest or head injury and no breathing difficulty, elevate your feet, legs and hips slightly above the level of your head. This will improve circulation to your brain and help you remain conscious.

Once you've treated yourself for shock, keep a close eye on your wound for resumption of bleeding. When the depressed effects of shock have been relieved, normal circulation returns to the blood vessels, and rapid bleeding may occur if the wound isn't sufficiently clotted and bandaged.

OTHER SELF-AID PRINCIPLES

For particular kinds of wounds and injuries there are a few more things that are good to know, beyond the 4 lifesaving steps.

Abdominal Wounds. Don't take anything—food or drink—by mouth. Don't try to replace intestines that are hanging out. Cover them with plastic or a moistened bandage. Put a pillow or other object under your knees to relax your abdominal muscles.

Neck Wounds. Don't use circular bandages around your neck—these may cut off your air and strangle you. Keep your head and shoulders raised. If you are stabbed or shot in the jugular vein or cut there by sharp glass or metal in a traffic accident, keep direct pressure on your wound all the way to the hospital.

Head Wounds. Keep your head elevated. Try not to bend your neck unnecessarily, because of the danger of neck fracture. Take nothing by mouth. Do *not* raise your feet. If blood or other fluid is coming out your nose, mouth or ear, turn your head to the draining side to make drainage easier.

Eye Injury. If a bullet, BB, knife, ice pick, or piece of glass, wood or metal penetrates your eye, don't try to remove it. Immobilize your eye by covering *both* eyes with a dressing (your eyes move together—if you're looking around with the uninjured eye, the injured eye will move, too). **Exception:** Don't cover both your eyes if you haven't brought a hostile threat under control. It's better to risk losing an eye than to blindfold yourself and let someone walk up and pump you full of lead.

Get to an eye doctor as soon as possible!

Fractures. Broken bones can grind around inside you and add serious new injuries. Spinal fractures can cause paralysis or death. If you suspect a fracture, stay as still as possible until help arrives. If you have to be moved, immobilize the fracture area by tying or strapping it to some inflexible object, such as a board, a piece of pipe, your shotgun, or your baton.

Internal Injuries. If you're pushed from a second-story balcony, or you impact with the steering wheel in a high-speed crash, or you take a direct shot to the chest, back or stomach where vital organs are in danger, you may suffer massive internal injuries. Take nothing by mouth. Remain still (seated; or lying on back). Turn to the injured side if necessary for drainage.

Foreign Objects. If you're walking around someone's backyard in the middle of the night on a prowler call and fall onto a wooden stake or a steel rod and become impaled, don't pull up off the object. Don't allow anyone to pull the object out of your body. Cut the object off outside your body, leave it imbedded, and have someone get you to the hospital.

Getting Help. There are some things you can't do for yourself, like artificial respiration, external heart massage, and, if your injuries are disabling, getting to the hospital. When you're injured in the daytime in a public place, getting help probably won't be a problem. But if you crash on a deserted country highway at midnight, or you're ambushed in a closed business district at 2 in the morning, you may have to do something to bring in help. If your partner is dead and you're disabled, use your *radio, lights,* and *noise* to attract attention.

If you can get to a working radio but you have trouble speaking up, place the mic directly against your throat and pronounce your words. This is no time for lengthy radio code, like "Unit Fourteen Bravo . . . I have a 10-97 . . . request assistance at 8th and Broadway." Just say: "Help. 8th and Broadway. Help."

If you can't use a radio, or if you need help **now** and the nearest ambulance or patrol car is 20 minutes away, you have to get the attention of passing motorists or nearby residents. Honk your horn with intermittent series of 3 short blasts. Use the high pitch of your siren to wake up dogs, who will wake up people. Flash your lights at oncoming cars. Blow your whistle. Wave flares. If necessary, throw a rock or your flashlite through the window of a closed store to set off the burglar alarm. If the situation requires it and you can safely do so, fire into the air.

When 1 untrained helper arrives, tell him how to call for assistance on your car radio, or once you're stabilized, send him to the nearest phone to call an ambulance. If 2 or more people arrive to help, select the most mature, calm, strong person to follow your first-aid instructions, and send the other person to call the ambulance. If there is a language barrier, don't try to talk in sentences—just say *words* like "Help. Ambulance. Police. Doctor. Help."

Shirt-Pocket Lifesaver. The first-aid kit in the trunk of your car is just fine when you're up and about and can get to it. But you won't have it handy if you're trapped in an overturned car, or shot in a back alley with your unit parked out on the street a death crawl away. And you probably haven't made a habit of carrying around an assortment of compresses, dressings, bands of cloth, plastic, and aluminum foil, because you've got all those other things to carry around on your Sam Browne belt.

If there is a single emergency medical supply item that I feel I can recommend to you without loading you down like a combat medic, it's the military triangular bandage. You can get one of these at an Army-Navy surplus store, or from your local National Guard or Reserve station, or Civil Defense quartermaster. It comes in a 2" x 4" packet, marked "Dyed Dressing, Bandage, Muslin, 37 by 37 by 52 inches." The packet is the same size as a pack of cigarettes, weighs only 2 ounces, and can be carried in your left shirt pocket, over your heart.

The triangular bandage can be used as a compress, dressing, tourniquet, sling, or tying device. It's large enough to use as a full chest or back bandage and can be folded into smaller widths for any size wound. Six diagrams on the back of the packet show you different ways to use the bandage. Inside the cardboard packet, the bandage is kept dry within a plastic pouch. This pouch can be cut or torn into a single, flat layer of plastic, 5" x 9", which you can use to seal a sucking chest wound or to cover intestines protruding from an abdominal wound.

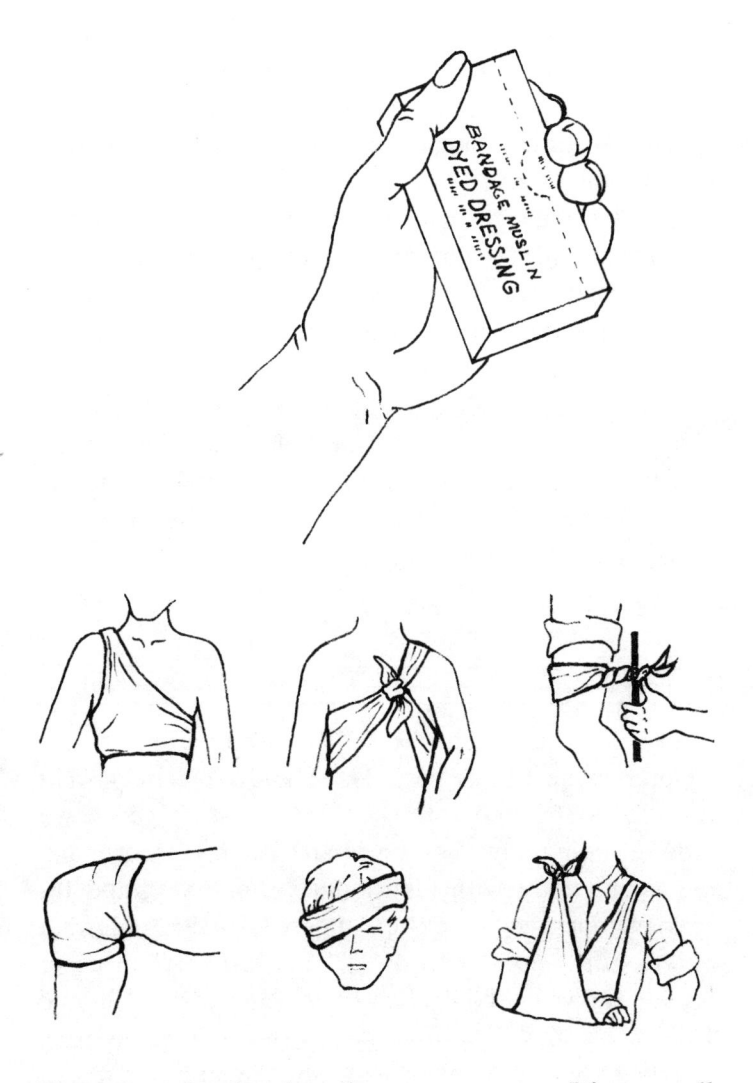

The triangular bandage has many uses. It's a small item of survival insurance.

Unlike the bright, white bandages you carry in your car's first-aid kit, the triangular bandage is dyed camouflage green. You can apply it and keep operating, if necessary, during a nighttime firefight, without becoming a walking target.

The triangular bandage is a *shirt-pocket lifesaver*. Don't leave it at home in your other uniform. Don't leave it in your locker. Don't leave it in your briefcase. Don't leave it in your patrol car. Even if you carry it in your shirt pocket for 20 years and retire without ever having opened the packet, I won't be sorry I recommended it to you. And if you pull it out some Friday evening and use it to save yourself from the next Monday's funeral roster, you won't be sorry you followed my recommendation.

RECAP

I've tried to keep the self-aid instructions in this chapter as simple as possible. I know precisely what it's like to be under fire and see a lot of your own blood running down your arm and hand, look over and see your buddy with his shoulder shot away, and try to remember what you're supposed to do, and what you're not supposed to do: "Should I set him up or lay him down? Should I give him water or not? Should I raise his feet?" If the answers don't come back clearly to you (and they may not if you don't review this chapter at least once a year), at least try to remember this much:

- *If you don't know whether or not you're supposed to do some particular thing,* **don't do it.**

- *If you can't remember which position is best,* **lie flat on your back.**

- *If you can't remember whether or not to eat or drink,* **don't do either.**

- *If you don't remember anything else about what you're supposed to do, think of the 3B's that you have to protect:* **blood, breath, and body heat.**

Give them something to work with when they get you to the hospital. Don't go in DOA. All you need to do is know how to save your life with self-treatment, keep calm, and do it. *You can survive* wounds and injury! □

NOTES ON LOCAL RULES

9

IT'S YOUR LIFE

Your survival is a personal matter. The decisions you make to insure it are personal. The practices you follow to protect yourself against your various occupational risks are up to you.

This book, like all other survival publications, contains no "right" and "wrong" concepts—only opinions and recommendations on concepts which *may* work, for *most* people, *most* of the time. It's up to you to compare survival alternatives, try them in practice, and decide for yourself what you think will work best *for you*. It's *your* life.

The proper position for coming home next Friday—and at the end of *every* shift, and one of the most important reasons why you should make SURVIVAL a personal goal.

STAY ALIVE ... *You can survive!* ☐

Order Today!
OUR MOST POPULAR CRIMINAL JUSTICE TITLES...

- ☐ ea. *California* Criminal Law $19.95
- ☐ ea. *California* Criminal Procedure $29.95
- ☐ ea. Community Relations Concepts $24.95
- ☐ ea. Courtroom Survival $14.95
- ☐ ea. Criminal Evidence $29.95
- ☐ ea. Criminal Interrogation $19.95
- ☐ ea. Fingerprint Science $19.95
- ☐ ea. The *New* Police Report Manual $14.95
- ☐ ea. The Officer Survival Manual $15.95
- ☐ ea. PC 832 Concepts $19.95
- ☐ ea. Police Unarmed Defense Tactics $9.95
- ☐ ea. Practical Criminal Investigation $29.95
- ☐ ea. Prin. of Amer. Law Enf. & Crim. Just. $34.95
- ☐ ea. Search and Seizure Handbook $15.95
- ☐ ea. Traffic Investigation and Enforcement $21.95
- ☐ ea. Understanding Street Gangs $19.95

Mail to:
CUSTOM PUBLISHING COMPANY
1590 Lotus Road Placerville, CA 95667
1-800-223-4838 *orders only*
(In California call (916) 626-1260)
Please enclose payment or departmental purchase order.

Card # ____ - ____ - ____ - ____   Visa/MC (circle one)

Exp.Date ___/___ Telephone(___)_____

Signature _____
(Signature required for all charge cards purchases)

Address _____

City _____ St ____ Zip _____
SALES TAX: California residents add applicable sales tax.

We Pay Shipping

Order Today!   *Prices Subject to Change.*

Order Today!
OUR MOST POPULAR CRIMINAL JUSTICE TITLES...

- ☐ ea. *California* Criminal Law $19.95
- ☐ ea. *California* Criminal Procedure $29.95
- ☐ ea. Community Relations Concepts $24.95
- ☐ ea. Courtroom Survival $14.95
- ☐ ea. Criminal Evidence $29.95
- ☐ ea. Criminal Interrogation $19.95
- ☐ ea. Fingerprint Science $19.95
- ☐ ea. The *New* Police Report Manual $14.95
- ☐ ea. The Officer Survival Manual $15.95
- ☐ ea. PC 832 Concepts $19.95
- ☐ ea. Police Unarmed Defense Tactics $9.95
- ☐ ea. Practical Criminal Investigation $29.95
- ☐ ea. Prin. of Amer. Law Enf. & Crim. Just. $34.95
- ☐ ea. Search and Seizure Handbook $15.95
- ☐ ea. Traffic Investigation and Enforcement $21.95
- ☐ ea. Understanding Street Gangs $19.95

Mail to:
CUSTOM PUBLISHING COMPANY
1590 Lotus Road Placerville, CA 95667

1-800-223-4838 orders only

(In California call (916) 626-1260)
Please enclose payment or departmental purchase order.

Card # ____ - ____ - ____ - ____ Visa/MC (circle one)

Exp.Date ___/___ Telephone(___)_____

Signature _____
(Signature required for all charge cards purchases)
Address _____

City _____ St ____ Zip _____
SALES TAX: California residents add applicable sales tax.

We Pay Shipping

Order Today!   *Prices Subject to Change.*